SpringerBriefs in Public Health

SpringerBriefs in Public Health present concise summaries of cutting-edge research and practical applications from across the entire field of public health, with contributions from medicine, bioethics, health economics, public policy, biostatistics, and sociology.

The focus of the series is to highlight current topics in public health of interest to a global audience, including health care policy; social determinants of health; health issues in developing countries; new research methods; chronic and infectious disease epidemics; and innovative health interventions.

Featuring compact volumes of 50 to 125 pages, the series covers a range of content from professional to academic. Possible volumes in the series may consist of timely reports of state-of-the art analytical techniques, reports from the field, snapshots of hot and/or emerging topics, elaborated theses, literature reviews, and in-depth case studies. Both solicited and unsolicited manuscripts are considered for publication in this series.

Briefs are published as part of Springer's eBook collection, with millions of users worldwide. In addition, Briefs are available for individual print and electronic purchase.

Briefs are characterized by fast, global electronic dissemination, standard publishing contracts, easy-to-use manuscript preparation and formatting guidelines, and expedited production schedules. We aim for publication 8-12 weeks after acceptance.

More information about this series at http://www.springer.com/series/10138

Daniel S. Goldberg

Public Health Ethics and the Social Determinants of Health

 Springer

Daniel S. Goldberg
Center for Bioethics and Humanities
University of Colorado Anschutz Medical Center
for Bioethics and Humanities
Aurora, CO, USA

ISSN 2192-3698 ISSN 2192-3701 (electronic)
SpringerBriefs in Public Health
ISBN 978-3-319-51345-4 ISBN 978-3-319-51347-8 (eBook)
https://doi.org/10.1007/978-3-319-51347-8

Library of Congress Control Number: 2017955829

Printed on acid-free paper

This Springer imprint is published by Springer Nature
The registered company is Springer International Publishing AG
The registered company address is: Gewerbestrasse 11, 6330 Cham, Switzerland

Contents

Chapter 1
Introduction

What causes disease? It seems the most simple of questions one could ask about disease from a public health standpoint. Yet there is nothing simple about it. Consider, for example, what causes the "white plague" of tuberculosis. The typical answer is "mycobacterium tuberculosis," and this is of course true in some sense. The mycobacterium is a necessary cause of the disease of tuberculosis. But is it also a sufficient cause of the disease? Does every carrier of the mycobacterium develop active tuberculosis? Indeed, does everyone exposed to the mycobacterium become infected?

Although it is tautological, public health is fundamentally public. This means that the unit of analysis is usually if not always groups and populations, a simple fact that, as we shall see, becomes enormously important for clear thinking in public health ethics. Thus, from the standpoint of disease causation, the interesting question from a population health perspective is not simply what causes tuberculosis in individual patients, but what drives *patterns* of tuberculosis in populations. And here, the evidence of stark inequalities in tuberculosis prevalence, incidence, and outcomes becomes significant. While mycobacterium obviously can be said to cause tuberculosis, it appears that other causal factors seem to determine which groups are most likely (1) to be exposed to the mycobacterium; (2) to develop active TB; and (3) to receive adequate treatment that can improve their outcomes. Moreover, as we shall see, the variables that determine the outcomes of a sickness go far beyond access to adequate medical treatment.

The interesting fact is that 19th c. public health reformers were well aware of the significance of understanding at what level we should think about the causation of public health. Consider, for example, a keystone document in the history of modern public health written by the physician, scientist, and anthropologist Rudolf Virchow entitled *A Report on a Typhus Epidemic in Upper Silesia* (1848). In this report, Virchow considered an outbreak of typhus in Upper Silesia, an economically depressed province in Prussia (and now Poland). Virchow's report carefully lays out the social and demographic context in which the epidemic occurred, and

© The Author(s) 2017
D.S. Goldberg, *Public Health Ethics and the Social Determinants of Health*,
SpringerBriefs in Public Health, https://doi.org/10.1007/978-3-319-51347-8_1

subsequently discusses the clinical course of typhus, differential diagnoses, case descriptions, and autopsy reports.[1] Moreover, as the first English-language translators of the Report point out, while "Virchow recognized the value of bacteriological research he could never accept a simply causal relationship between bacterium and disease."[2] After describing some of the political, social, and economic difficulties in which Silesians found themselves in 1848, Virchow put it simply: "There cannot be any doubt that such a typhoid epidemic was only possible under these conditions, and that ultimately they were the result of the poverty and underdevelopment of Upper Silesia. I am convinced that if you changed these conditions, the epidemics would not recur."[3]

Virchow proposed as a solution to the typhus epidemic and to prevention of similar outbreaks a broad agenda of social reform: "In theory, the answer to the question as to how to prevent outbreaks in Upper Silesia is quite simple: education, together with its daughters, freedom and welfare."[4] More specifically, Virchow suggested a number of proposals including universal education, democracy, tax reform, and agricultural/industrial development.[5]

The justification for these proposals, and the all-important question of whose responsibility their fulfillment might be are central issues in this book. But for now, all that matters is understanding how Virchow's 'prescription' turned on his belief in the fundamentally social causation of the typhus epidemic. Indeed, only a few sentences later, Virchow explicitly shows that the lens through which he is viewing disease and health is fundamentally collective and social, rather than training his focus on the individual clinical case: "Thus, for us, it is no longer a question of the medical treatment and care of this or that person taken ill with typhoid, but of the well-being of one and a half million citizens who find themselves at the lowest moral and physical decline."[6]

Virchow's analysis here underscores another important point in thinking about the causation of disease: to what extent is access to health care services—or its absence—a primary determinant of population health outcomes? And to what extent are differences in those outcomes between groups driven by access to medical care? Virchow seems to be suggesting in the above quote not that medical care for the sick is unimportant, but that the social and political factors he views as primary causes of the typhus epidemic shape the pattern of disease across a large population. If this is correct, it immediately implies that, at the level of the relevant population, the most important remedies are targeted at ameliorating the social and political problems that cause the outbreak.

Have we succeeded in complicating the "simple" question of 'what causes disease?' In truth, the philosophy of disease causation is vastly more complex than this

---

[1] Rieger & Taylor 1985, p. 550.

[2] R&T, p. 550.

[3] Ibid., 551.

[4] Ibid., 551.

[5] Ibid., 550.

[6] Ibid., 551.

brief introduction suggests, but for purposes of opening up a discussion in public health ethics and the social determinants of health, merely asking and then engaging the question is extremely helpful. Although this book is not a philosophical exposition of disease causation, it takes as its point of departure that epidemiologic evidence that addresses the question of 'what are the primary factors that cause patterns of disease in human populations?' Because epidemiologists and public health scientists are generally well-aware that questions of disease causation are extraordinarily complicated, the evidence base tends to focus on "determinants" of health and disease rather than "causes." This book follows the general trend, and speaks primarily of "determinants" rather than "causes" of health.

Although this brief book is fundamentally a book about public health ethics, the ethical analyses are not possible without an adequate grounding in what the evidence actually suggests regarding the connections between socioeconomic conditions and population health outcomes. Accordingly, Chap. 2 addresses The Epidemiologic Evidence Regarding the Social Determinants of Health. The chapter considers the historical and contemporary evidence suggesting that social and economic conditions are in fact the prime determinants of patterns of disease in human populations. Although it is difficult to prove a negative, the chapter also considers the converse question: among possible contenders, which factors does the evidence suggest are not chief determinants of population health? And given that the determination of health is not an all-or-nothing affair—different factors can contribute more or less to overall health outcomes—the question of relative contribution (how much do different factors contribute to overall health outcomes relative to each other) will also be addressed in Chap. 2.

Note that virtually all of the questions and concerns discussed in Chap. 2 are empirical or descriptive questions. That is, they are questions about the way the world is. We cannot answer these questions with ethical or normative inquiry, or analysis of the way the world ought to be. Before we can move on to engage the ethical questions, the state of the evidence must be fleshed out to form an adequate foundation on which ethical analysis can rest. Of course, the fact that we must engage empirical epidemiologic evidence does not imply that such evidence is settled or beyond question. Indeed, a host of questions, challenges, and controversies surround the epidemiologic evidence base regarding the social determinants of health, and some of these issues will be raised and highlighted in Chap. 2 where appropriate.

Chapter 3 begins the book's exploration of some of the key ethical implications of the evidence regarding the social determinants of health. Specifically, the chapter focuses on three concepts that are crucial both to public health ethics in general and more specifically to ethical analysis of the social determinants of health: justice, compound disadvantage, and health inequities. These concepts are intimately related, and the chapter will provide a thorough grounding in the meaning of these important concepts and the connections between them.

Although "justice" is a notoriously difficult philosophical concept, most modern analyses include in their discussion some reference to the notion of desert: What do people in society deserve? What are they owed? And who is responsible for satisfying which obligations to which people? It is not possible to discuss moral obligations to improve population health without addressing complicated questions of responsibility for health outcomes. But note that here again, the level of analysis matters greatly; individual responsibility for health might differ in morally important ways from collective responsibility for health. Moreover, responsibility for an individual person's health might also differ in morally important ways from responsibility for a group's health.

Chapter 4 takes up these questions in earnest and applies them to a public health matter of grave importance: the role of risky health behaviors in determining population health outcomes. Here too, important social and moral questions abound, such as:

- To what extent can individuals or groups be said to be responsible for such behaviors?
- What does "responsible" mean in this context?
- What evidence supports possible attributions of responsibility?
- How do notions of responsibility for risky health behaviors alter our notions of what people deserve (i.e., notions of justice)?

Chapter 5 steps back from some of the specific concerns and issues discussed in Chaps. 3 and 4 to consider their potential application specifically at the level of public health policy and practice. Many of the most significant ethical ramifications of the evidence regarding the social determinants of health fall under the general question of priority-setting. Chapter 5 will cover ethical questions of priority-setting at the level of policy and practice—whether global, national, regional, local, or hyperlocal—to discern the extent to which the evidence regarding social determinants of health complicates difficult ethical questions regarding priorities in public health. For example, even if stakeholders agreed that compressing health inequities is a vital public health goal, the question of 'which health inequities' matter most would remain to be resolved. Thinking about these intricate questions in terms of the primary ethical objectives of public health policy is central to public health ethics.

The priorities set by public health policy drive practice, and no text on ethics and the social determinants of health is complete without scrutiny of its ramifications for public health practice. Therefore, Chap. 5 addresses the extent to which the evidence regarding the social determinants of health ought to determine public health practice at the local levels in which such practice is often situated. What can individual public health practitioners do about adverse socioeconomic conditions impacting health? Is it within the purview of such practice to address such conditions and intervene using tools and techniques internal to public health practice?

Finally, the Conclusion synthesizes some of the principal themes and concepts addressed in the core of the book, and lays out directions and questions through which to frame further study on the subject of ethics and the social determinants of health.

How to Use This Book

This book blends together features of a scholarly monograph and a textbook. Monographs are typically although not always intended for an audience of peers, whether active academics or advanced students. By contrast, textbooks often accompany a student's introduction to a particular topic, and help set the stage for subsequent inquiries that presume a sufficient level of background knowledge. This book integrates features of both texts. Each chapter begins with a general discussion of the chapter topic, the purpose of which is to introduce the reader to some of the central content, questions, and debates regarding that topic. This introductory discussion helps scaffold the second portion of each chapter, which offers a variety of exercises and discussion questions that guide the reader to application of the substantive chapter content to central issues in public health policy and practice.

Chapter 2
Ethics, Justice, and the Social Determinants of Health

As noted in the Introduction, in order to evaluate the ethical implications of the issue at hand, the epidemiologic evidence base regarding the social determinants of health must be laid out in sufficient detail. This second chapter offers a broad overview of some of this evidence in order to lay the groundwork needed to draw out many of the key ethical issues that arise in evaluating and ordering public health priorities based on the social determinants of health.

To begin, we can again return to the basic question briefly examined in the Introduction: what causes disease? Or, to phrase the question in the terminology that this book adopts, 'what are the prime determinants of health and its distribution in human populations?' In the U.S. at least, the evidence suggests that most people tend to think of health as a function of access (or lack thereof) to health care services. But is this correct? For purposes of understanding the evidence base regarding the social determinants of health, a basic distinction between health and health care must be clearly understood. While health care services can certainly have moral value even if they do not produce health, it is difficult to deny that the overarching goal is improvement in health outcomes rather than delivery of health care services. Indeed, health care services which have been shown not to improve health are generally considered wasteful or inefficient and are disfavored from a health policy perspective.

And the policy perspective is extremely important, not simply in this chapter, but for the book as a whole. Especially at higher levels, policy obviously affects large numbers of people. So while an individual health care provider may or may not have a good clinical justification for using an intervention for which the evidence of safety and efficacy is weak, recommending widespread adoption as a matter of public health policy is inadvisable.

Accordingly, to understand the evidence base regarding the social determinants of health, it is necessary to begin with a distinction between health and health care services. This book presumes that the primary moral goal is improvement in population health rather than any increase in access to health care services. Such a

© The Author(s) 2017
D.S. Goldberg, *Public Health Ethics and the Social Determinants of Health*,
SpringerBriefs in Public Health, https://doi.org/10.1007/978-3-319-51347-8_2

presumption certainly does not deny that an increase may in fact be good or morally obligatory. But the primary goal of public health policy and practice ought to be improvement in population health. What is the standard by which we can judge whether a given intervention improves population health?

As noted in the Introduction, defining health turns out to be an enormously difficult and complicated inquiry. Fortunately, these conceptual problems do not preclude us from being able to operationalize health so as to be able to measure and assess health status at any given point of time, as well as changes in such status in response to specific variables. Epidemiologists use a variety of different measures to assess the health status of a population, but the most generic assessments are simply morbidity and mortality (each of which of course contains a number of different component sub-values depending on the particular variable being investigated).

There is another important criterion for population health: the extent to which different groups within the overall population have worse health outcomes than other groups. Public health authorities at almost any level are in general agreement that, other things being equal the greater the inequalities in important health outcomes within a given population, the worse that population's health can be said to be. Accordingly, for this chapter and for the book as a whole, we can adopt what has been referred to as the two primary goals of ethically optimal public health policy:

1. improvement in overall population health;
2. compression of health inequities.

For convenience, we can refer to the first criterion as the Absolute Health prong (because it refers to absolute improvements in health) and the second criterion as the Relative Health prong (because inequalities in health are by definition relative to group health status). Do not worry if these terms do not make perfect sense now; Chaps. 3 and 4 are devoted to explaining the significance of these ideas of absolute and relative health.

While the idea that improvement in overall population health is a primary ethical goal is not likely to attract significant dissent, the same cannot be said for the second goal. Explaining why many, although not all public health ethicists deem health inequities a serious moral problem, and hence their contraction an important moral good, is the subject of Chap. 3. For now, it suffices to understand that in general, the criteria for improvement in population health are *both* improvements in overall population health (typically measured by outcomes such as mortality and morbidity) and the compression of health inequities.

As implied in the Introduction, one of the best ways of understanding the significance of the social determinants of health is to turn to history. This is true with regard to analysis of the evidence base as well. We can begin with a physician and demographer named Thomas McKeown. Beginning in the early 1950s and carrying though the 1970s, McKeown and colleagues examined over four centuries of birth and death records meticulously compiled in local parishes in England. The rich data enabled McKeown and his collaborators to analyze the single largest recorded gain in life expectancy in the known world. Between approximately 1581 and 1945, life expectancy in England more than doubled, an astonishing gain in a relatively short amount of time.[1] (In truth, most of the gains in life expectancy occur between 1700 and 1945, an even shorter time period). Indeed, if there are larger increases across a relatively large population in such a short time period, we do not have records that show it.

Naturally, one of the most important questions from an epidemiologic perspective is 'what were the prime determinants of this singular gain in life expectancy?' Given that the period in question covered both the Industrial and the so-called Therapeutic Revolutions, it is obviously imperative to understand what factors seem most responsible for the remarkable health improvements during the time. McKeown et al.'s analysis of precisely this question resulted in multiple papers, several books, and no small amount of controversy that continues to this day. For our purposes, the ensuing McKeown Thesis can be boiled down to two basic claims:

1. What we think of as modern allopathic medicine had almost nothing to do with the massive increase in life expectancy; and
2. Organized public health had a greater but still very small causal effect on the life-years gained during the relevant time period.

The argument for the first claim is relatively straight-forward. The first broad class of drugs that met anything like current standards of safety and efficacy were the sulfa antibiotics, which began to be manufactured in the 1930s. McKeown and his colleagues showed convincingly that virtually all of the diseases responsible for the bulk of English mortality were in substantial decline well before the sulfa drugs were manufactured and made available to the general population. For example, McKeown et al. showed that the death rate from tuberculosis was already in decline before Koch identified the bacillus responsible, and, moreover, that most of the declension in the curve occurs well before the first effective chemotherapeutics, let alone the advent of vaccination.

The inescapable conclusion of the McKeown group's analysis is that modern medicine had very little to do with the single largest increase in life expectancy in recorded history. Instead, McKeown and colleagues argued, improvements in nutrition and rises in the general standard of living were responsible for these stunning population health improvements. (For purposes of our discussion of the McKeown Thesis, we are only discussing the Absolute Health prong; we are not yet discussing health inequalities). McKeown's group also denied that the rise of organized public health activity in the mid-to-late 19th and early 20th c. had a substantial impact on the sharp decrease in English mortality observable from the 1830s and 1840s through the 1940s and 1950s.

What is especially important about the McKeown Thesis is the fact that, while the second claim—that public health had little to do with the increased life

---

[1] For a summary of the Mckeown Thesis, see James Colgrove, "The McKeown Thesis: A Historical Controversy and Its Enduring Influence," *American Journal of Public Health* 92, no. 5 (2002): 725–729.

expectancy—is extremely controversial, the first claim—that organized medicine had little do with the observed gains—is simply not. As leading historian of public health Simon Szreter puts it,

> [The McKeown Thesis] effectively demonstrated that those advances in the science of medicine forming the basis of today's conventional clinical and hospital teaching and practice, in particular the immuno- and chemo-therapies, played only a very minor role in accounting for the historic decline in mortality levels. McKeown simply and conclusively showed that many of the most important diseases involved had already all but disappeared in England and Wales before the earliest date at which the relevant scientific medical innovations occurred.[2]

The McKeown Thesis, then, is an important piece of data that underscores the need to distinguish between health and health care in thinking about the prime determinants of health and its distribution. For the record, McKeown's second claim regarding the general ineffectiveness of organized public health in accounting for the mortality declines has held up less well, and Szreter has been among those leading the charge in arguing that public health policy and practice had a substantial impact in the absolute health improvements.

The McKeown Thesis is in part an historical argument, but it is one that stretches well into the 20th c. The obvious question, then, is whether evidence of the more recent past and the present supports the conclusion that health care services, however important they are in caring for the sick, is likely not a major determinant of health and its distribution. In addition, while the McKeown Thesis helps us discern what is likely *not* a principal determinant of health, it is more controversial and arguably less helpful in illuminating what *are* such chief determinants.

Fortunately, contemporary evidence sheds light on both of these important issues. One of the best sources of evidence with which to continue our investigation are the Whitehall Studies, led by Sir Michael Marmot. Marmot, a physician and social epidemiologist in the United Kingdom, inaugurated a longitudinal study on the health of British civil servants in the 1960s, the follow-up for which is ongoing today. The quality of the data used in the Whitehall Studies is nothing short of remarkable.[3] The original study included over 18,000 participants, and tracked them longitudinally over a significant period of time. Relatively few participants dropped out of the study, and because British society features relatively rigid class structures, the investigators were able to control for a number of confounders. Doing so enabled the research group to isolate the relationship between employment grade and mortality. The original hypothesis for Whitehall I was that those participants working at the highest employment grade would have the most stress and thereby have higher mortality risks. But the study confirmed exactly the opposite.

What the evidence showed is an almost stepwise increase in mortality from cardiovascular and heart disease as the employment grade decreases. Those at the highest level of employment experience the lowest mortality, while those at the lowest

---

[2] *Health and Wealth: Studies in History and Policy* (Rochester: University of Rochester Press), 99.

[3] Gopal Sreenivasan, "Health Care and Equality of Opportunity," *Hastings Center Report* 37, no. 2 (2007): 21–31.

level (below clerical) experience the highest mortality. Moreover—and this is extremely important—those employment grades neither highest nor lowest also demonstrated this stepwise increase in mortality. That is, Marmot and colleagues found a robust correlation between employment grade and mortality at every level of the social hierarchy, not just at the top and the bottom.[4]

Employment grade is actually a reasonably good proxy for socioeconomic status ("SES"), and the approximately linear relationship between SES and mortality in a given population has come to be referred to as the *social gradient of health*. The evidence supporting the idea of a social gradient in health in different communities, at different population levels, in entirely different places around the globe is nothing short of immense. As Marmot has observed, we find a social gradient in health virtually everywhere we care to look.

One of the best metaphors for understanding the social gradient of health is a ladder. Indeed, in one of the Whitehall Studies, participants were invited to mark their self-perceived place in the social hierarchy by using a ladder.[5] Those who marked the lowest status were much more likely to report their health as poor or fair than those who marked higher social statuses.

While the outcome here is self-reported health, it is worth noting that self-rated health is considered a fairly reliable health indicator, and has been found to correlate reasonably well with objective population health indicators of morbidity. In other words, while many people may not have a particularly good understanding of pathophysiology or the mechanisms of disease, many people are in fact able to assess whether they are generally healthy or not.

So, one of the first crucial pieces of epidemiologic evidence needed to understand the ethics of the social determinants of health is the social gradient of health: absolute health outcomes tend to be strongly and robustly correlated with SES or a related but not-identical measure known as "socioeconomic position" ("SEP"). Moreover, the social gradient itself self-evidently represents the state of health inequalities in a given population; each rung of the ladder features a group with a different health status than members of the group at higher/lower rungs. And long-term follow-up of the subject participants in Whitehall II showed a similar social gradient of health, which indicates that the relationship between employment status and mortality is robust (i.e., it persists over time).

There are two other important points we can draw from scrutiny of the Whitehall Studies. First, the participant population in the studies has since the inception of Whitehall I been British civil servants. Through the National Health Service, British civil servants enjoy universal access to at least basic health care services. It follows that the stark differences in mortality found across the participant population very likely is not due to differences in access to health care. This reinforces the central

---

[4] Michael G. Marmot, Geoffrey Rose, M. Shipley, and P.J.S Hamilton, "Employment Grade and Coronary Artery Disease in British Civil Servants," *Journal of Epidemiology & Community Health* 32, no. 4 (1978): 244–249.

[5] Robert Sapolsky, "Sick of Poverty," *Scientific American* 293 (2005): 92–99.

negative point of the McKeown Thesis, that access to medical care is generally not a prime determinant of health and its distribution in human populations.

Second, Marmot and colleagues were interested in the extent to which clinical risk factors could explain the mortality outcomes they found. Even combining the effects of hypertension, smoking, and cholesterol did not account for more than 1/3 of the total mortality burden experienced by the group at the lowest employment grade. Marmot and colleagues concluded that risk factors and risky health behaviors alone cannot account for the social gradient of health.

If you find this last point difficult to believe, you are not alone. A number of people do not agree with the Whitehall research group that risk factors are likely not a primary explanation for the social gradient of health.[6] Moreover, once we begin to think about risky health behaviors, we open up space to consider a number of crucial moral issues relating to desert and responsibility, issues that will be taken up in earnest in the next chapters.

Thus far in this chapter, we have considered evidence drawn from largely two sets of research: the McKeown analysis and the Whitehall studies. Are the general findings as described here supported by other epidemiologic evidence? The answer is "overwhelmingly, yes." Both the negative and positive theses discussed thus far in this chapter are supported by so much evidence that they are often taken as a given among communities of scholars and practitioners who are generally informed on what has been taken to being called "social epidemiology" (which focuses on "the effects of socio-structural factors on states of health").[7]

The evidence supporting both of these theses is synthesized admirably in the 2008 final report of the World Health Organization's Commission on Social Determinants of Health (chaired, not coincidentally, by Sir Michael Marmot), entitled *Closing the Gap in a Generation: Health Equity through Action on the Social Determinants of Health*. In Chap. 2, the report notes that societies have "traditionally looked to the health sector to deal with its concerns about health and disease." And while unequal distribution to health care services are both empirically and ethically important—a claim we shall consider in Chap. 3—"the high burden of illness responsible for appalling premature loss of life arises in large part because of the conditions in which people are born, grow, live, work, and age …."[8] These conditions collectively can be termed the social determinants of health.

In an important 2007 article, Lantz, Litchtenstein, and Pollack decry what they term as the "medicalization of health policy," which is indeed a problem if health is not mostly a function of access to medical care.[9] They point out the irony in the fact that "inequalities in access to health care are often smaller than corresponding

---

[6] Silvia Stringhini et al., "Association of socioeconomic position with health behaviors and mortality. The Whitehall II study," JAMA 303, no. 12 (2010):1159–1166.

[7] Kaori Honjo, "Social epidemiology: Definition, history, and research examples," *Environmental Health & Preventive Medicine* 9, no. 5 (2004): 193–199.

[8] WHO Report, p. 26.

[9] "Health Policy Approaches to Population Health: The Limits of Medicalization," Health Affairs 26, no. 5 (2007): 1253–157.

inequalities in access to housing, education, nutrition, and other resources."[10] The latter are "often more important than personal health services in generating or ame- liorating health inequalities."[11] In short, "lack of access to health care is not the fundamental cause of health vulnerability or social disparities in health."[12] Lantz, Lichtenstein, and Pollack's analysis here addresses both the Absolute Health prong and the Relative Health prong of our criteria for health. Health status is largely a function of determinants such as housing, education, SEP—the conditions in which people "are born, grow, live, [and] work"—as are differences in health status between groups (i.e., health inequalities).

Other than the above, time and space do not permit us to examine the immense amount of evidence supporting the idea that social and economic conditions are prime determinants of health and its distribution in human societies. But, again, the state of the evidence is nothing short of overwhelming on this point, so much so that it is not seriously disputed in general, and certainly not within public health contexts in particular. One final way of framing this evidence comes to us from medical sociologists Bruce Link and Jo Phelan, who in 1995 titled an important paper in a way that underscores the discussion in Chap. 2: "Social Conditions are Fundamental Causes of Disease."[13] Although the theory in many ways simply reiterates the evi- dence base above, it provides a useful framework for examining some of the ethical implications of said evidence base, and is therefore worth examining. Moreover, as noted in the Introduction, thinking about disease causality is critical to evaluating these ethical implications, and thus even where identifying determinants is more practical than sussing out causation, perspective on causality is unavoidable.

It is no coincidence that, as quoted above, Lantz, Lichtenstein, and Pollack state that "lack of access to health care is not *the fundamental cause* of health vulnerabil- ity or social disparities in health." A given factor qualifies as a fundamental cause when it satisfies three basic criteria. First, the factor must cause multiple diseases. Second, the factor must determine multiple risk factors. Third, the factor must per- sist over time. As a prime example of a fundamental cause, consider SES. Given the evidence regarding the social gradient of health, we know beyond a shadow of a doubt that SES is associated with a number of diseases, both infectious and non- communicable. Second, we know that the distribution of risk factors in the U.S. population, for example, is highly sensitive to SES. Thus, the lower down the social gradient one goes, the higher the smoking prevalence and incidence.[14] The same is generally true for many other so-called risky health behaviors.[15] This epidemiologic fact—that risk factors such as health behaviors often track social gradients in

---

[10] Ibid., 1256.

[11] Ibid.

[12] Ibid.

[13] *Journal of Health and Social Behavior* Spec. Issue (1995): 80–94.

[14] Centers for Disease Control, "Fact Sheet: Health Disparities in Cigarette Smoking," available at http://www.cdc.gov/minorityhealth/chdir/2011/factsheets/smoking.pdf

[15] Institute of Medicine Committee on Health and Behavior, *Health and Behavior: The Interplay of Biological, Behavioral, and Societal Influences* (Washington: National Academies Press, 2001).

general—has hugely important ethical implications that will be discussed at length in upcoming chapters.

In any event, SES also satisfies Link and Phelan's third criterion for fundamental causes; its effects persist over time. Link and Phelan turn to the history of public health to make the point: in the 19th c., one of the major risk factors for disease was access to adequate sewerage. Because the affluent tended to enjoy better access, SES shaped this specific risk factor, and the poor bore a disproportionate burden of waterborne disease. Gradually, access to sanitation became more equal, and such access ceased to be a major population-wide risk factor for disease. Yet other mechanisms (i.e., tobacco consumption) have arisen that reflect the tight connection between SES and disease, showing the persistence needed to establish that SES is in fact a fundamental cause of disease.

Of course, fundamental cause theory is hardly the only framework through which to view the connections between socioeconomic conditions and population health. In many ways, it serves simply to reinforce some of the larger themes and ideas of the epidemiologic evidence base discussed in this chapter.

Conclusion and Summary

By now, the question with which we began has been answered, or at least has been answered to the extent needed to go on and consider some of the relevant ethical implications. What causes disease in human populations? Or, given our interest in avoiding some of the complexities of discerning disease causation, what are the prime determinants of health—and its distribution—in human populations? The epidemiologic evidence is quite clear regarding both the negative and the positive answers to this latter question. The negative answer—what is *not* such a prime determinant—is important because of the general tendency in the U.S. at least to conflate health and health care, to see population health as a general function of access to health care services. Abundant historical and contemporary evidence suggests that this widespread belief is unsupportable; while important for a variety of reasons we shall consider in Chap. 3, access to health care services is nevertheless likely only a relatively minor determinant of population health.

The positive answer to the above question suggests that social and economic conditions are the prime determinants of health and its distribution in human populations. Or, to put it in the language of the WHO's Commission on Social Determinants of Health, the conditions in which we live, work, and play are the chief determinants of health. Moreover, defining health for purposes of this book requires understanding health outcomes both in terms of their absolute status and their relative distribution within the relevant population (hence the Absolute Health prong and the Relative Health prong of our working definition of health).

Equipped with some of these foundational understandings of the epidemiologic evidence regarding the social determinants of health, we can now move on to assess some of the key ethical issues that such evidence raises.

Discussion Questions

1. The data set for the McKeown Thesis runs roughly from 1581–1940. Is it fair to apply conclusions drawn from 17th, 18th, and 19th c. records to present arguments about the chief determinants of health? Why or why not?
2. Why does it matter whether access to health care is a prime determinant of health and its distribution?
3. Explain the distinction between absolute and relative health. Why does it matter?
4. Are you inclined to agree with the original finding in Whitehall I that risk factors cannot account for the social gradient in health? Why or why not? If clinical risk factors are not major determinants of unequal health status, what do you think might be driving the inequalities observed in the Whitehall Studies?
5. Do you think health policy in the U.S. tends to be medicalized? If so, is this a problem? Why or why not?

Chapter 3
Justice, Compound Disadvantage, and Health Inequities

In Chap. 2, we reviewed the epidemiologic evidence suggesting that social and economic conditions are prime determinants of health and its distribution in human populations. The reference to "distribution" is important, for it underscores the dual criteria used in thinking about health improvements in this book. As we shall see in this chapter, it is insufficient for a public health intervention to simply improve absolute health in a given population. Rather, there is general consensus that compressing health inequalities is a crucial moral goal of public health practice and policy, and therefore how health is distributed within a population matters for thinking about ethical priorities in public health.

Accordingly, this chapter, which begins the normative ethical analysis of the book, is focused on social and health inequalities. Such emphasis does not imply that the Absolute Health prong is somehow less significant than the Relative Health prong (i.e., that improvements in overall population health are morally less important than compression of health inequalities). Part of the power of the evidence surveyed in Chap. 2 is that it shows a strong connection from socioeconomic conditions to both Absolute and Relative Health. There is, of course, little ethical dissent from the idea that improvements in Absolute Health are morally important. More importantly, the ethical basis for such a belief is readily apparent to most: public health interventions that improve the overall health of a population is a morally good consequence. And while the fact of good consequences across a large number of moral agents is not the end of moral inquiry, the idea that such a state of affairs is in at least some sense morally desirable seems difficult to contest.

That health inequalities are morally problematic is arguably less intuitive to most, and requires some analysis, which is the focus of this chapter. There is almost uniform agreement among public health ethicists that discussing distributions of health and their moral implications centers on the principle of justice. The meaning of the term "justice" is unfortunately extraordinarily complex. Accordingly, trying to unpack and explain some key features of justice in context of health inequalities and the social determinants of health is the central goal of this chapter.

© The Author(s) 2017

D.S. Goldberg, *Public Health Ethics and the Social Determinants of Health*,
SpringerBriefs in Public Health, https://doi.org/10.1007/978-3-319-51347-8_3

Although it might be a bit of a stretch to claim that public health ethics is synonymous with "justice," there is little question that the principle is front and center in almost any public health ethics analysis. This emphasis does not imply the exclusion of other important values, of course (i.e., nonmaleficence, autonomy, etc.), but it does immediately show an important distinction between medical ethics and public health ethics. The widely-used "four principles" approach to medical ethics (autonomy, beneficence, nonmaleficence, and justice) unquestionably centers the principle of patient autonomy. This focus is unsurprising given the special and traditional ethical significance afforded the physician-patient relationship. That is, medical ethics has often focused on the individuals most prominent in a therapeutic relationship, most often characterized as the health care provider and the patient.

As noted in both the Introduction and in Chap. 2, public health ethics is fundamentally public; its problems and interventions typically occur on the group level. Changing the unit of analysis from individual patients to groups and populations dramatically alters the ethical implications, and requires different concepts, emphases, and tools for ethical analysis. It is, therefore, a grave mistake to think that ethical analysis in public health can simply proceed by applying traditional medical ethics analyses to the level of groups and populations. Precisely because of their individualistic focus, traditional approaches to medical ethics map poorly onto public health ethics. Indeed, presuming that properties of individuals automatically apply to properties of groups—even where groups are aggregates of individuals—is a kind of error often referred to as the fallacy of composition or the fallacy of division.

Groups behave differently than individuals, and analyzing ethical problems that adhere to publics is in important ways different from analyzing ethical problems that arise from individual clinical treatment encounters. This is not to suggest that there is no overlap, of course, but merely that the kinds of ethical analyses we shall need to use to think about the social determinants of health should not begin from the same starting point as that used for thinking about, e.g., whether life-sustaining treatment may be withdrawn from a particular patient.

The epidemiologic evidence discussed in Chap. 2 grounds a basic understanding of the ways in which social and economic conditions produce health within and across a given population. It follows that differences in those baseline social and economic conditions would be expected to produce different health outcomes in populations differentially exposed. That is, if one population were consistently exposed to adverse social and economic conditions, and a different population were consistently exposed to supportive and nurturing social and economic conditions, we would predict differences in health between the two groups.

This is exactly what we find. There are stark differences in health outcomes among and between groups at almost any level of analysis, whether hyperlocal, local, county, regional, state, national, global. The idea that social and economic conditions are prime determinants not simply of absolute health outcomes, but of differences in health outcomes between groups, is the prime descriptive premise that grounds ethical analysis. The question we now must ask is whether these facts about the world are ethically problematic.

Are all health inequalities morally wrong? If not, which? How do we distinguish between those that are morally tolerable and those which are not? Which health inequalities are of the highest moral priority? Does thinking about "justice" help us grapple with these questions? Why or why not?

At the outset, we should note that the existence of inequality is not necessarily morally problematic. For example, there is a stark inequality in the basketball talent of LeBron James as compared to that talent possessed by the author of this book. Should we be concerned morally about this state of affairs? Obviously not. Even as applied to health inequalities, a plausible argument suggests that some group-level differences are not inherently morally troubling. Consider the pronounced gender differences in prostate-cancer screening. Leaving aside for the moment complicated questions over the propriety of such screening in general, it is obvious that the fact of gender-based inequalities in prostate-cancer screening is entirely copacetic. We should expect such differences where, generally, women are extremely unlikely to be diagnosed with prostate cancer—the predictable result of generally not possessing prostates!

Yet other kinds of health inequalities intuitively strike us as morally troubling. For example, a baby born tomorrow in Sweden has a life expectancy of approximately 82. A baby born tomorrow in Sierra Leone has a life expectancy of approximately 45. This is a mind-bogglingly large discrepancy. While we often attribute differential health outcomes to willing participation in risky health behaviors (see Chap. 5) we do not have recourse to this move here by virtue of the fact that the subjects in question are infants. Although we might want to preserve at least the possibility that a near 40-year-difference in life expectancy between Sweden and Sierra Leone can be morally justified, our intuitions tell us that this fact is morally suspect. (Moral philosophers typically argue that intuitions alone are insufficient to ground a moral claim, but such intuitions are nevertheless a good indicator of whether a particular issue is morally problematic or not. In other words, our intuitions are good places to begin a moral analysis, although we cannot stop with them).

To push the point a bit farther, we might ask what it is about the Sweden-Sierra Leone life expectancy gap that strikes us as morally troubling? The most likely explanation is that it feels somehow unfair that an infant born tomorrow in Sweden is likely to live nearly forty years longer than an infant born tomorrow in Sierra Leone. But why? What does "unfair" mean in this context? And what is the criterion for fairness at all?

Questions of fairness open up discussions of justice. Indeed, Plato himself defined justice as fairness, as giving each person his/her due. Plato obviously understood that phrased as such this definition of justice is something of an empty vessel—indeed, his work is largely his effort to fill that container with content—but the framework he provides is extremely important. We can, at a very broad level, think about justice as a question of desert: what do we owe each other? What distribution of goods, services, and outcomes is fair in a just social order?

Scholars have engaged Platonic conceptions of justice for thousands of years. We cannot hope for an in-depth, sophisticated analysis of the truly immense body of work on the subject in this book, let alone in this chapter. What follows here, then, is a birds-eye-view of three influential conceptions of justice that are generally thought to be of particular significance for thinking about population health in general and of the social determinants of health in particular. The discussion here must not be taken as exclusive; a variety of other approaches, frameworks, and models of justice have been fruitfully explored and would undoubtedly be worth applying in thinking through some of the ethical implications of the evidence regarding the social determinants of health.

Rawls and Distributive Justice

It is virtually impossible to mention theories of justice without discussing John Rawls's famous 1971 book *A Theory of Justice*. It is no exaggeration to suggest that in the four decades since its publication, any serious attempt to think through the philosophical implications of any model of justice must engage with and account for Rawls's theories, even if the aim is dissent. Rawls's analysis focused on distributive justice, i.e., what distribution of important goods is fair? To answer this question, Rawls developed a famous thought experiment he termed the "veil of ignorance." He asked his audience to imagine a hypothetical state of nature (i.e., a world in which a formal society with norms and laws had not yet formed) in which none of the participants knew in advance what social status they would hold in the future society. He called this the "Original Position." Behind this veil of ignorance, the participants would have to choose how they would distribute important goods, without the benefit of knowing the shares of such goods that they would receive in their society. Rawls argued that this thought experiment illuminated the criteria for a fair distribution of goods, consisting primarily of two principles of justice:

1. all members of the society must be guaranteed basic equal rights and liberties needed to "secure fundamental interests of free and equal citizens and to pursue a wide range of conceptions of the good."
2. all members of the society must receive a fair distribution of "educational and employment opportunities enabling all to fairly compete for powers and prerogatives of office;" and securing "for all a guaranteed minimum of the all-purpose means (including income and wealth) that individuals need to pursue their interests and to maintain their self-respect as free and equal persons."

Rawls did not believe that all inequalities are necessarily unjust. Indeed, it is probably not too much of an exaggeration to suggest that the question of "which inequalities are unjust" is one of the central questions that animates discussions of justice, including as to health status and its social determinants. Rawls's second principle of justice actually allows for the existence of some inequalities. Note that he does not argue, for example, that all members of society must be guaranteed the same *outcomes*; rather, his emphasis is on equality of certain opportunities, including the chance to participate in politics and exert an influence in a democratic social order.

But, he continued, the only permissible inequalities would be those which would benefit the least advantaged (Rawls called this "The Difference Position"). That is, social and economic inequalities are only consistent with justice to the extent that those inequalities "make the least advantaged class better off than they would be in any other feasible economic system ...." It is worth pausing here for a moment to evaluate the extent to which most Western democracies adhere to this requirement—do most inequalities in the developed world confer the greatest benefit on the least advantaged? Regardless of your answer, what do you think are some moral implications?

Of course, how these principles apply to the social determinants of health is not obvious. Intriguingly, Rawls had very little to say about health or health care in his book, although many have since applied Rawlsian analysis to health. Among the most influential of these expositors is Norman Daniels, whose work is relatively unique insofar as he has explored the implications of Rawls's approach to the distribution of *health* rather than simply as to the distribution of *health care*. (This point obviously underscores the tendency noted in the book to focus on health care rather than health!) In his 2008 book, *Just Health: Meeting Health Needs Fairly*,[1] Daniels sets out what he terms the Fundamental Question: "As a matter of justice, what do we owe each other to promote and protect health in a population and to assist people when they are ill or disabled?"[2] Note that this formulation tracks the broad rule-of-thumb suggested above for thinking about questions of justice (roughly, "what do we owe each other?)" Yet Daniels argues that this question is almost impossible to answer as written, and suggests that it is more fruitful to engage three Focal Questions that, when answered, provide sufficient guidance on the Fundamental Question:

1. "Is health, and therefore health care and other factors that affect health, of special moral importance?"
2. "When are health inequalities unjust?"
3. "How can we meet health needs fairly under resource constraints?"

As to the first question, Daniels claims that health is indeed of special moral importance because is required to protect "fair shares of the normal opportunity range" that is in turn necessary for human functioning. As to the second question, health inequalities are unjust to Daniels when it "is the result of an unjust distribution of the socially controllable factors affecting population health and its distribution." (Of course, this answer merely puts to Daniels the task of explaining what it is about particular distributions of such factors that render them unjust). As to the third question, Daniels argues that when health needs are calibrated and satisfied according to what he terms "accountability for reasonableness," health needs are being met fairly. The important point here is that Daniels's framework is procedural. That is, we are meeting health needs fairly when we show accountability for reasonableness, which in turn has four criteria: (1) publicity; (2) relevance; (3) appeals; and (4) enforcement.

---

[1] Cambridge: UK (Cambridge University Press).
[2] Ibid., p. 11.

Satisfying these four criteria literally constitutes "meeting health needs fairly," which means that following the process demanded by Daniels's framework is the way in which we fulfill our obligations to each other (in context of health). The procedural nature of Daniels's analysis underscores a major tension in thinking about what we owe each other in context of health, and how the social determinants of health bear on this question. Many theorists argue that although we may reach widespread agreement on some of the broad parameters of what we owe each other in ameliorating adverse social and economic conditions, no account of justice can really indicate what specifically we must do to intervene in any given situation. The formal way of phrasing this well-known problem is to say that 'models of justice under-determine action guidance.' That is, no matter which model of justice we prefer, such models are, often enough, not going to provide useful and explicit answers regarding the specific choices that public health actors may face.

This is obviously a large problem for any attempt to provide substantive answers for resolving thorny questions of desert in context of the social determinants of health. If theories of justice cannot guide our actions with regard to improving population health and compressing health inequalities, what use are they? Tracking Rawls, thinkers like Daniels argue that the best we can do—and it is quite a lot!—is to ensure that we have a robust process of what some call "public reason," by which members of society have fair and equal opportunities to decide as a society what action we are obligated to undertake in terms of the social determinants of health.

Although most commentators working on justice and health tend to agree that the concept, as crucial as it is, generally may not provide specific guidance on any particular public health problem, it obviously does not follow that the concept of justice is empty. That is, defining justice and giving the term substance matters even if that substance may not in all cases determine the right action. While Rawlsian approaches to thinking about justice have arguably dominated over the past few decades, several important criticisms have sprung up during this time, one of which comes from the political economist and philosopher Amartya Sen.

The Capabilities Approach

Beginning in the early 1980s, Sen developed a theory of justice that came to be called the capabilities approach.[3] Along with its other principal expositor, the philosopher Martha Nussbaum,[4] the capabilities approach typically begins with a criticism of the Rawlsian emphasis on the distribution of primary goods. In thinking about what we owe each other, Sen argued, it is insufficient to focus simply on how social goods are distributed, because a life of human flourishing is not reducible to such goods. Resources like income and education are crucial, but they are, as philosopher Ingrid Robeyns puts it, "particular means to well-being rather than the

---

[3] See Amartya Sen, *The Idea of Justice* (Cambridge: Harvard University Press, 2009).

[4] *Creating Capabilities* (Cambridge: Harvard University Press, 2011).

ends."[5] Robeyns explains that the "capability approach prioritizes certain of peoples' beings and doings and their opportunities to realize those beings and doings (such as their genuine opportunities to be educated, their ability to move around or to enjoy supportive social relationships)."[6]

Determining what we owe each other is not simply a function of how we allocate income, because "people differ in their ability to convert means into valuable opportunities (capabilities) or outcomes (functionings)." What matters most, in Sen's evocative language, is the freedom "to do and be." Means to those ends are important, but, according to Robeyns, "means can only work as reliable proxies of people's opportunities to achieve those ends if they all have the same capacities or powers to convert those means into equal capability set."

Even among moral and political philosophers, the capabilities approach is notoriously difficult to pin down. Yet it has proven extremely popular and influential especially in global development theory and practice, and is therefore of particular importance in thinking about global health. The most notable effort to apply the capabilities approach to a population health context comes from the philosopher and social theorist Sridhar Venkatapuram. In his 2011 book, aptly entitled *Health Justice*, Venkatapuram sets out to give an account of justice based on the capabilities approach that is squarely concerned with health.

Even more explicitly than Daniels's work, Venkatapuram integrates the social epidemiologic evidence base discussed in Chap. 2 into his account. He dispenses relatively quickly with the emphasis on the distribution of health care services (one favored by a long tradition of Rawlsian analyses, therein demonstrating how commonly health and health care are conflated), and argues instead that the central criterion for health justice is the capability to be healthy: "a person's ability to achieve or exercise a cluster of basic capabilities and functionings, and each at a level that constitutes a life worthy of equal human dignity in the modern world." Unlike Rawls, who had little to say about health outside of health care, Venkatapuram contends that the extent to which a society supports the capability to be healthy for its constituents tells us a huge amount about its commitment to health justice.

For Venkatapuram, the significance of the capability to be healthy springs from the idea of what it is to be human at all, to be a "needy temporal animal being." Accordingly, the capability to be health is what he terms a "pre-political moral entitlement." Societies obviously have an enormous role to play in ensuring the satisfaction of this capability, but political orders do not create it.

You have probably noticed that neither Daniels nor Venkatapuram makes much reference to the idea of "social justice." This is not accidental. Although the term is often used in conversation among public health academics and leaders, philosophers and theorists have long pointed out that the concept is often extremely ill-defined in practice. Critics, in fact, have sarcastically argued that the concept of "social justice" is

---

[5] Robeyns, Ingrid, "The Capability Approach", *The Stanford Encyclopedia of Philosophy* (Summer 2011 Edition), Edward N. Zalta (ed.), URL = <http://plato.stanford.edu/archives/sum2011/entries/capability-approach/>

[6] Ibid.

something of an empty vessel, which can often be interpreted as referring to whatever conception of fairness and rightness the particular speaker happens to have in mind.

If the concept of social justice is to have content, and, more importantly, if it is to persuade anyone who does not already share a speaker's perspective, it cannot be reduced to a mere slogan. To be sure, a community of public health practitioners may share a number of moral commitments, and, in this capacity, reference to "social justice" may simply reflect that overlapping consensus. Such a use of the concept is not without value, but if the goal is to understand how and why the evidence regarding the social determinants of health compels particular acts, practices, and policies, reflection of shared perspectives is by itself insufficient. This reflection does not offer any account by which stakeholders and policymakers could identify and order policy priorities or decide which inequalities matter most.

Social Justice and Public Health Ethics

Bioethicists Madison Powers and Ruth Faden argue that far from being a mere slogan, the concept of social justice actually stands as the moral foundation of public health and health policy.[7] Their theoretical framework adopts what they term a "sufficiency" approach to defining social justice. Under Powers and Faden's account, what we owe each other is determined not by a distribution of primary goods through a veil of ignorance, nor via the capabilities that a just social order ought to provide for its constituents. Rather, justice is determined by the extent to which a society provides a basic minimum of resources sufficient to promote well-being across six essential dimensions of well-being.

Akin to Venkatapuram, Powers and Faden's account suggests that certain minimal factors are absolutely necessary for any human being to flourish. These factors are not relative to different societies and cultures, but are, at a very broad and minimal level, universally required. Powers and Faden term these the six dimensions of well-being:

1. Health
2. Personal Security
3. Reasoning
4. Respect
5. Attachment
6. Self-determination

Powers and Faden's account, like Daniels's and Venkatapuram's analysis, integrates the social epidemiologic evidence base discussed in Chap. 2.[8] They argue

---

[7] *Social Justice: The Moral Foundations of Public Health and Health Policy* (New York: Oxford University Press, 2006).

[8] In fact, virtually any leading theory of justice developed by public health ethicists is based on this evidence base, which both underscores the importance of the latter in framing ethical analysis in public health and justifies the approach taken in this book.

that, in thinking through how a sufficient level of health can be achieved, the single most importance piece of evidence is the fact of compound or clustered disadvantage. We are already familiar with this idea from the discussion in Chap. 2. Again, the basic idea here is that social disadvantages tend to cluster in groups. If you belong to a group that experiences one kind of social disadvantage, you are significantly more likely to experience additional social disadvantages as well—as a member of that group. Thus, if you belong to a group that as a group tends to experience low socioeconomic position ("SEP"), you are also as a member of that group likely to experience low educational attainment, substandard and/or hazardous housing, and greater exposures to violence, racism, and stigma. Of course, just because a group experiences one of the social disadvantages does not guarantee that it experiences any single other kind of disadvantage. Nor does the fact that disadvantage clusters at the group level mean that *all* individuals within that group will experiences *all* of the disadvantages observable at the aggregate level.

But, across the group in question, the clustering of disadvantage generally holds, and most of the members of that group will likely experience the effects of compound disadvantage. Powers and Faden deem this fact immensely important, and develop an evocative phrase for the problem: the creation and sustained existence of "densely-woven patterns of disadvantage." Powers and Faden use the metaphor of a web to characterize one of the key features of the pattern: "Inequalities of one kind beget inequalities of another, and over the course of a lifetime … the compounding of disadvantage makes avoidance or escape difficult without heroic effort or unexpected good luck."[9]

This is an important passage both within Powers and Faden's account and for understanding the ethical implications of the social determinants of health. First, the quotation observes the empirical fact of compound disadvantage: social inequalities seem to feed off of themselves and each other—they cluster. Second, the quotation has a reference to the "course of a lifetime." This underscores the significance of a life course approach to understanding the social determinants of health, and of the way in which adverse conditions especially early in life can not only have important health effects later in life, but can produce such effects in subsequent generations.

But life course epidemiology has enormously important ethical implications as well. In the first place, insofar as interventions geared towards the social determinants of health can have special impact on conditions in early childhood, they may improve Absolute Health not only for those affected, but for those people's offspring in the next generation. Indeed, the evidence supporting both the efficacy and the cost-effectiveness of intensive early childhood development ("ECD") programs is extremely impressive. Nobel laureate and economist James Heckman has researched the subject extensively over the last decade, and on the basis of the evidence has developed what he terms The Heckman Equation:

---

[9] Ibid., 193.

The Heckman Equation

Invest	in educational and developmental resources for disadvantaged families to provide equal access to successful early human development.
+Develop	cognitive skills and social skills in children — from birth to age five when it matters most.
+Sustain	early development with effective education through adulthood.
=Gain	more capable, productive and valuable citizens that pay dividends to America for generations to come.

Among the robust evidence supporting ECD, one of the most notable findings comes from The Carolina Abecedarian Project, a randomized and controlled longitudinal study conducted in North Carolina examining the link between such ECD and a variety of social outcomes (including health). A recent paper tracking the Project noted (as the title of the paper) that "Early Childhood Investments Substantially Boost Adult Health," documenting "significantly lower prevalence of risk factors for cardiovascular and metabolic disease."[10] Moreover, and as we shall see as we further explore Powers and Faden's theory, it is important when discussing the social determinants of health to avoid the idea that health is all that matters in thinking about what we owe each other. We know that social and economic conditions exert an enormous effect in producing distributions of health in human populations, but there is more to human flourishing than health. This point is important to Powers and Faden's account of social justice, and we shall return to it momentarily. In terms of the Carolina Abecedarian Project, however, what matters is that the study showed that a variety of outcomes important to overall well-being *other than health* are robustly connected to intensive ECD. These included significant improvements in measures of employment, educational attainment, and age at the birth of the first child.

Moreover, because cost-benefit ratios for The Abecedarian Project are estimated at 2.5:1, the kinds of intensive ECD envisioned by the Project are highly cost-effective, a finding borne out to greater or lesser extent from the evidence on ECD in general. It is worth pausing here to ask the following question: "What if ECD did not save money?" That is, what if ECD did improve a variety of important social outcomes but cost more money than it saved? Does it follow that investment in ECD is morally impermissible? Morally permissible? Morally mandatory?

---

[10] Frances Campbell et al., "Early Childhood Investments Substantially Boost Adult Health," *Science* 343, no. 6178 (2014): 1478–1485.

The first possibility, that investment in ECD is morally impermissible, seems intuitively implausible. It takes little imagination to think of activities in which we are morally permitted to engage which end up costing more money than the benefits are worth. Few would suggest, for example, that even if maintaining emergency departments costs much more money than it is worth (and its worth can obviously be weighed in a variety of ways) that we are morally prohibited to intervene in emergent situations to save someone's life. The third possibility also seems implausible, at least depending on the extent to which any given intervention is cost-negative. Because money is a scarce resource, funds allocated to one intervention are funds not expended on a different possible intervention, and, if the amount of money needed to intervene in a given way dwarfs the value of the benefits obtained, it is difficult to see why doing so would be mandatory in all cases.

The second possibility is the most intuitively acceptable one—it seems plausible to assert that even where interventions are not cost-saving or cost-effective, they may still be morally permissible. An easy example supporting this proposition in a population health context is that of smoking cessation. Many people believe that smoking cessation is cost-saving. But in truth, this is not the case. The reason why smoking cessation results in higher lifetime health expenditures is that people who smoke tend to have much lower life expectancy than non-smokers. When smokers quit, their life expectancy increases, and so do lifetime health expenditures. Furthermore, because the vast majority of health expenditures in the U.S. occur in the last 6–12 months of life (and are higher in older populations), smokers who quit and live at or past average U.S. life expectancy (74) end up costing more health expenditures than smokers who, as a group, tend to die relatively quickly and relatively young (compared to non-smokers).

The evidence underlying this surprising result—that quitting smoking ends up costing society more money than not quitting—actually holds across a number of prevention paradigms and risk factors, and is the basis for what has been termed "the prevention paradox." The important point here is that the mere fact that cessation of smoking is not cost-effective rather obviously does not imply we are morally prohibited from encouraging such cessation. Most people would agree that quitting smoking is a social good, one that should be encouraged and in which resources should be invested, even if it results in higher lifetime health expenditures.

This is an important point in public health ethics in general. Too often the value of efficiency or cost-effectiveness is simply assumed to be the primary moral good. While there is little doubt that efficiency in general is "a" social and moral good, its primacy must not be assumed. There are obviously cases in which economically inefficient public health interventions may be morally permissible—even morally mandatory in some cases—and the argument must be made rather than the conclusion simply being assumed.

Although cost-effectiveness is obviously relevant to Powers and Faden's theory of social justice, it does not exhaust the inquiry. And as to ECD, the fact that it is cost-effective is only one point in its favor. The fact that investment in ECD likely has intergenerational impact is another point. Social epidemiologist Hilary Graham has argued that, as a rule, public health officials have paid far too little attention to what he terms "future publics."[11] The fact that at least some public health interventions (e.g., climate) can have fairly dramatic impacts on future generations is morally significant. Deciding the weight given to such benefits is an enormously complicated question, and much of it depends on the extent of our possibly different moral obligations to what Daniels terms identified vs. statistical victims.[12] Do we owe more to a person standing in front of us than to a person who will not be born for 25 years? Why or why not? Nevertheless, the fact that early childhood is an epidemiologically sentinel period is morally significant, even if determining how those impacts are to be cashed out in terms of moral value remains an open question.

This fairly detailed discussion of ECD is important insofar as it touches on a number of issues central to assessing the ethical implications of the evidence regarding the social determinants of health. In terms of Powers and Faden's theory, one of the most significant points, as mentioned above, is the fact that investment in ECD seems to improve outcomes in a number of important areas other than health. This matters because Powers and Faden are at pains to reject what might be termed "health exceptionalism." Namely, that while health is of course important, is not the only social good. People often trade health security and status for other ends that they wish to pursue, and Powers and Faden readily acknowledge that people may reasonably seek a number of ends in pursuit of the good life even where those ends may result in a trade-off or sacrifice of health.

Nevertheless, Powers and Faden do believe that there are some features so basic to a life of human flourishing that they are essential for any person, regardless of how they choose to order their life. Powers and Faden term these the six essential dimensions of well-being:

1. health
2. personal security
3. reasoning
4. respect
5. attachment
6. self-determination

The criteria for social justice is the extent to which a society provides to its members the social conditions necessary for its members to achieve a *sufficient* amount of these six essential dimensions of well-being. The theory places a lot of weight on the idea of sufficiency, so much so that Powers and Faden's theory of justice has been termed a "sufficientarian" account. So, how do we know whether a given health intervention comports with mandates of social justice? If the intervention helps to create or sustain conditions that enable people to obtain sufficient amounts of the six essential dimensions of well-being, it satisfies Powers and Faden's framework of social justice.

---

[11] "Where is the Future in Public Health," *The Milbank Quarterly* 88, no. 2 (2010): 149–168.

[12] Norman Daniels, "Reasonable Disagreement about Identified vs. Statistical Victims," *The Hastings Center Report* 42, no. 1 (2012): 35–45; *Identified vs. Statistical Lives* (eds. I. Glenn Cohen, Norman Daniels, and Nir Eyal) (New York: Oxford University Press, 2015).

The fact that Powers and Faden's model of social justice is based on sufficiency highlights the importance of densely-woven patterns of disadvantage. For, as Powers and Faden point out, those caught within these webs or clusters of disadvantage are disproportionately likely to experience insufficient amounts of the six essential dimensions of well-being. That is, people belonging to groups that are subject to compound social disadvantage are less and less likely to be able to achieve a sufficient level of health or personal security or a sufficient opportunity to determine the course of their own lives (self-determination). This claim in turn is at least partly supported by the epidemiologic evidence regarding the clustering of disadvantage and its impacts on stress and health. But it should not take too much imagination to understand how, for example, people who are persistently exposed to violence (a form of social disadvantage) are in the aggregate less likely to achieve sufficient levels of personal security.

Because compound disadvantage poses immense difficulties for health sufficiency, it becomes a priority in Powers and Faden's theory. One of the most difficult questions with which theorists of justice grapple is ordering ethical priorities. This problem is of particular importance to public health; some argue that priority-setting is the central ethical question in the entire field of public health ethics. Priority-setting issues are extremely common in public health, in multiple ways and at multiple levels. For example, when a state government cuts appropriations to local health departments, the leadership of the latter face difficult decisions regarding how to revise and realign their priorities in the face of less resources. And in pandemic or disaster planning, it is crucial to determine which public health workers should enjoy priority in receiving prophylactic antibiotics or antivirals, not to mention which patients are of highest priority for receiving different kinds of interventions.

But, and as we will see in Chap. 5, issues of priority-setting go beyond the individual cases mentioned above. Priority-setting is also a vital component in determining what problems, policies, and categories of interventions are favored and implemented at global and domestic levels. Public health leaders, for example, face difficult choices in deciding whether to prioritize interventions that address public health problems higher up the causal pathway (often referred to as upstream, distal, or structural factors) or that operate lower down the pathway (downstream or proximal factors). In other words, public health actors may be called upon to decide whether to allocate scarce resources towards interventions that seek to change upstream social determinants of health, or towards interventions targeted further downstream, closer to the onset of disease itself (i.e., clinical screening).

Arguably, a key task for any viable framework of public health ethics is the extent to which it can be applied in practice. Therefore, although it may be interesting and useful from time to time to think about what a utopian world would look like, many commentators agree that viable theories of justice should have meaning and relevance to the messy, nonideal world in which we live. In this, public health ethics is analogous to public health practice itself, which, as will be discussed in chapters four and five, must often proceed in the face of imperfect evidence. Accordingly, many accounts of health justice, including the health sufficiency approach of Powers & Faden and the capabilities approach as articulated by Amartya Sen, Martha Nussbaum, and Sridhar Venkatapuram, expressly avow a nonideal approach. This matters because it hopefully increases the likelihood that such theories may have utility in guiding actual practice. Moreover, nonideal theories of justice are also more modest in scope than attempts to prescribe universal accounts of right and wrong that govern in all contexts. Admittedly, some people think that this modesty is a mistake, and that any valid account of justice in public health must be universally applicable.

Powers and Faden, however, do not believe that this is correct. They acknowledge that their own account of justice underdetermines action guidance, a notion mentioned above.[13] Even after one correctly applies their theory to a given problem, a number of competing morally plausible alternatives may still exist. Accordingly, a theoretically solid nonideal theory of justice may still leave public health actors holding the bag, so to speak. Powers and Faden see this as an inevitable consequence of the messiness of the world; even the best moral theories are not algorithms that spit out the correct moral answers. It does not follow, of course, that nonideal theories in public health ethics are useless. Quite the contrary, there seems little question that theories which provide some moral guidance, or give good reasons for discarding morally insufficient policies and practices, are valuable. Even if we are unsure whether to select between options A, B, and C, knowing that options D-J are ethically unacceptable is valuable.

Health Equity

Much of the book thus far has emphasized the close connections between health inequalities and the social determinants of health. Many leading public health entities and actors have rallied around the concept of "health equity" as a tool for galvanizing action on the social determinants of health. Influential public health scientists such as Sir Michael Marmot and Paula Braveman expressly endorse the concept, as does internationally significant public and global health organizations such as the Centers for Disease Control and the World Health Organization. Curiously, however, public health ethicists are not as optimistic on the significance of the concept. Instead, a widespread if not quite unanimous view among practicing public health ethicists is that the idea of "health equity" is vague to the point of incoherence. Consistent with Plato's conception of justice, which defines justice as fairness, we are informed that health inequities which are avoidable are unfair and, hence, unjust. "Avoidability" is therefore a central criterion for health inequity (Preda and Voigt 2015).[14] But this criterion is problematic for a variety of reasons. For example,

---

[13] Powers and Faden, 32–35.

[14] Adana Preda & Kristin Voigt, "The Social Determinants of Health: Why Should We Care?" *American Journal of Bioethics* 15, no. 3 (2015): 25–36.

[g]ender-based inequalities in prostate cancer screening are in some sense avoidable. But the fact that we could, if we chose, compress such inequalities in screening does not make such inequalities unfair or unjust; quite the contrary, the obvious intuition is that such gender-based disparities are perfectly morally tenable.[15]

Moreover, Preda and Voigt note that the concept of "avoidability" tends to conflate two distinct sense of what is avoidable: what can be prevented, and what can be addressed. This matters, they explain, because "many unpreventable inequalities are in fact amenable to intervention,"[16] by some medical interventions. Preda and Voigt also note that under several leading conceptions of health equity, what makes avoidable health inequities unfair is the lack of control people have over the source of such inequities. Although this seems intuitively correct, the problem is that many natural inequalities are also uncontrollable, which means that such natural inequalities should be unjust for the same reason as social inequalities. Ultimately, Preda and Voigt argue that "avoidability" is neither necessary nor sufficient to explain what is unjust about health inequalities. Moreover, they are not alone in their concern that the idea of health equity is vague and undertheorized.[17]

Two questions are relevant here: first, does this matter? And second, if it does, what are the implications? If public health ethics matters at all to public health practice and policy, it is difficult to explain why problems in a leading moral concept said to ground public health action is irrelevant. One of the reasons public health ethics is important is its ability to help public health stakeholders justify particular positions, policies, and interventions. If the justifications for concepts deemed central to the entire field of public health are confused, inadequate, or unclear, they are unlikely to convince anyone who does not already agree with the speaker!

And yet, even confused or inadequate concepts may be politically important if they spur positive public health action. "Health equity" and its criterion of avoidability has

"served as a banner around which public health actors around the globe rally to address the devastating health inequalities that persist at almost every level of social organization. There is little dispute among ethicists that these inequalities are of enormous moral concern. Thus, it is plausible to assert that whatever any theoretical shortcomings in the idea of health equity, where the concept itself helps galvanize and organize public health action devoted to a top priority in public health practice and policy, it has tremendous social and political value.[18]

Ultimately, the debate over the concept of health equity and its significance for public health remains contested. Arguably, the most important point is for readers and public health stakeholders to understand that while "health equity" is currently at the center of practice and policy within the public health community, there is much that remains unclear about its meaning and its power to ground and justify any particular public health intervention.

---

[15] Daniel S. Goldberg, "The Naturalistic Fallacy in Ethical Discourse on the Social Determinants of Health," *The American Journal of Bioethics* 15, no. 3 (2015): 58–60.

[16] Preda: Voigt, p. 30.

[17] E.g., Wilson, J. *Health inequities*. In Public health ethics, ed. A. Dawson, 211–230. (Cambridge, UK: Cambridge 2011).

[18] Daniel S. Goldberg, "On the Very Idea of Health Equity," *Journal of Public Health Practice & Management* Supp. 1 (2015): S11–S12.

Conclusion

In this chapter, we have explored how social and economic conditions, among other structural factors, create webs of compound disadvantage that exert an enormous impact on population health outcomes. Although the persistence, characteristics, and distribution of such factors are epidemiologic facts about the world, it should by now be clear that these facts have profound ethical implications for public health practice, policy, and priority-setting. Social disadvantages are aggregate or group-level phenomena. Even individual members of significantly disadvantaged groups can enjoy more or less of those disadvantages than other members of the same group. Yet, this raises an important moral question, one touched on but not fully engaged in this text yet: to what extent do individuals bear moral responsibility for their own health outcomes? The fact that social determinants have a powerful role in shaping population health does not imply that individual acts and behaviors have no influence on health outcomes, especially at the individual level. And while public health is fundamentally about publics or groups, we know that patterns of behavior appear at the group level as well as at the individual level.

Exploring some of these questions and configuring the ethical significance of individual behaviors for public health practice and policy in light of the strong evidence regarding the socialization of health is the subject of the following chapter.

Discussion Questions

1. Why do traditional approaches to medical ethics distort good ethical analysis of public health problems?
2. How do we distinguish between health inequalities that are morally tolerable and those which are not?
3. Of the three concepts of justice discussed in this chapter, which do you find the most compelling? Why?
4. What does it mean to say that "models of justice under-determine action guidance?" Why does it matter for public health practice?
5. What is "compound disadvantage"? Why does it matter ethically?
6. Identify at least one ethical implication of the evidence drawn from life course epidemiology.
7. Why is Powers and Faden's account of social justice termed a health sufficiency model?
8. Why is health equity a contested concept? Does it matter that public health ethicists find the idea problematic? Why or why not?

Chapter 4
Ethics, Responsibility and Social Patterning of Risky Health Behaviors

Debates over individual responsibility for health outcomes are nothing new in American society. Historian Howard Leichter observes that "[t]he 20th century began and ended with many of the nation's health policymakers and opinion shapers blaming individuals for their own ill health."[1] If, for example, individuals have knowingly behaved in ways that damage their own health, it is unclear what obligation society owes that individual to meliorate the resulting bad health outcomes.[2]

Yet, the evidence regarding the social determinants of health belies any simple assertion of individual responsibility for health. The reason for this is that social and economic conditions that exist well beyond any one individual's agency are primarily responsible for the distribution of health in human populations. Individual acts almost certainly play some role in producing health outcomes across a population—although, as we shall see, the extent of that role remains vigorously disputed. Nevertheless, it is crucial to recall the point made at the outset of this book regarding the difference between moral analysis of individual vs. group units. That individual acts impact health outcomes for *that* individual is trivially true. An individual decision to fire a nail gun into the palm of one's hand will obviously impact the actor's health. Yet, the fallacy of composition bars us from moving directly from analysis of individuals to analysis of groups. The obvious fact that individual acts exert a strong impact on that individual's health does not license the claim that individual acts exert a strong impact on the health of the group as a whole.

Considerations of such responsibility are not merely academic; they are codified into health policies within both the U.S. and the U.K. Debates over individual responsibility for public health and the connection between assessments of moral culpability and public health policy are especially resonant in the U.S. This is

---

[1] "'Evil Habits' and 'Personal Choices': Assigning Responsibility for Health in the twentieth Century," *The Milbank Quarterly* 81, no. 4 (2003): 603–626.

[2] Howard Wikler, "Personal and Social Responsibility for Health," *Ethics & International Affairs* 16, no. 2 (2002): 47–55.

© The Author(s) 2017

D.S. Goldberg, *Public Health Ethics and the Social Determinants of Health*,
SpringerBriefs in Public Health, https://doi.org/10.1007/978-3-319-51347-8_4

because there is little question that the political culture in the U.S. lionizes individualism, a peculiarity that distinguishes the U.S. from many of its comparators in the West.[3] Accordingly, while questions of individual responsibility for health may be relevant everywhere, in the U.S., assessments of that culpability takes on outsized importance in shaping the politics of public health. While public health practitioners may or may not participate in such political discourse, they are in no way insulated from such discussions, which have an enormous impact on what kinds of public health practices and interventions are supported and prioritized.

This chapter surveys some of the literature on individual responsibility for health primarily through an examination of the connection between behaviors and public health. The epidemiologic evidence shows, perhaps surprisingly to some, that individual health behaviors cannot be understood outside of their social contexts. That is, there is good evidence that, where social and economic conditions impact health, they act within and through health behaviors. In other words, health behaviors do not exist in a vacuum; just as hard population health outcomes, they exist in a variety of patterns and distributions that reflect broader social and economic conditions. We shall discuss the evidence in section II of this chapter; the first task is to assess some of the reasons why the extent of individual responsibility for health matters morally.

Why Individual Responsibility for Health Matters

Many people intuit that individual responsibility for health matters, but how and why it matters turns out to be surprisingly complex. Harald Schmidt, one of the leading scholars of personal responsibility and health, notes that the phrase "X is responsible for P" in context of health can have both prospective and retrospective meanings.[4] As to the former, the phrase can mean that

- X should do P because no one else can do P for X;
- X should do P because P will be good for the health of X;
- X should do P because justice demands this.[5]

As to a retrospective meaning, Schmidt observes that assertions of individual responsibility typically mean any of the following:

1. that X has played a causal role in bringing about P;
2. that X has played a causal role in bringing about P, should recognize this, and should try to avoid doing so in the future;
3. that X has played a causal role in bringing about P, should recognize this, should try to avoid doing so in the future, and should pay back any costs; and

---

[3] The 2012 American Values Survey, available at http://www.people-press.org/values/
[4] "Just Health Responsibility," *Journal of Medical Ethics* 35, no. 1 (2009): 21–26.
[5] Ibid., 22.

4. that X has played a causal role in bringing about P, should recognize this, should try to avoid doing so in the future, should pay back any costs, and may be given a lower priority than patients whose behavior played a lesser role in contributing to their health outcomes

Although unpacking the full complexity of the claim "X is responsible for P" is not possible in this chapter, in terms of public health ethics, we can focus on item 4 in particular. 4 claims several obligations owed by those who have played a causal role in bringing about an adverse health condition. Beyond recognition and avoidance, agents who have brought about such health conditions are obligated to pay back any costs expended on their behalf, and may also be deemed a lower priority for response than those whose acts have not played any such role in their own adverse condition. Although not noted in 4, we might also, as bioethicist Dan Wikler notes, "insist that the potential risk taker pay in advance for insurance against added risk, either in the form of a user fee or a specific task. One example is the user fee for dangerous sports, intended to cover the added costs of paying for care in case of accident."[6]

We have already seen how questions of priority-setting are core to understanding the ethical implications of the social determinants of health. So, under 4, we might, for example, claim that people who have damaged their livers due to excessive and long-term alcohol consumption are less deserving of a liver transplant than one who requires a new liver as the result of contracting hepatitis due to a contaminated blood transfusion. But note that, to many people, the legitimacy of this controversial claim depends at least in part on volition, on the extent to which a person has control or influence in causing the adverse health condition ("P" in 4 above). We can see this when we note how important it is for many people to state that alcoholism is a disease, or when others claim that obesity is due to a genetic condition. The rhetoric here seems to suggest that the adverse health condition is not *caused* by any act or behavior of the individual in question, and therefore that none of the possible obligations laid out in 4 are applicable.

Of course, even if an individual unquestionably plays a causal role in bringing about an adverse health condition, such as the person who consumes excessive quantities of alcohol over decades, we might well have reason to deny the obligations and penalties described in 4. We will consider these arguments in a bit more detail in the next section; for now, it is important simply to understand why claims of individual responsibility for health might matter, and how they might be connected to underlying social disadvantage and inequality. That is, we know beyond a shadow of a doubt that more disadvantaged groups are in aggregate significantly less healthy than more powerful, affluent groups (i.e., the social gradient of health). Yet, if the root causes of those health disadvantages are deemed to lie within the risky health behaviors of the worst-off, 4 suggests reason for thinking that marginalized groups might deserve *lower* priority when allocating resources intended to improve Absolute Health.

---

[6] Wikler, at p. 113.

Note how significant this claim is for the framework we have developed thus far. The Relative Health prong of our overall conception of public health ethics suggests that compressing health inequalities is a crucial criteria for justice itself. But, as Wikler puts it, "the claim that inequalities represent injustices might be [open to revision] if the blame for the poorer health status of the less-favoured groups can be located in the individuals themselves."[7] In other words, one might question the entire basis for the Relative Health prong—that we should "level up" the least well-off in an effort to compress health inequalities—if the least well-off are deemed responsible for their unequal health status. Moreover, we need not argue that the least well-off should be cut off entirely from social assistance to undermine the Relative Health prong of our framework. 4 does not claim that individuals who have played a causal role in bringing about their health conditions are entitled to no public resources of any kind. Rather, it materially obligates such individuals either to defray or recompense the public for expenditures made on their behalf owing to the relevant adverse health condition(s), or to accept a lower priority in accessing scarce social resources.

Wikler frames the point this way:

> The locus of blame is key, for if blame is placed on the individual, social structure is exculpated, and the resulting suffering and premature death will not be counted as a social injustice. Narrowing health inequalities among social groups would thus not be of special urgency, either as a matter of prevention or of remedy.[8]

Thus far, we have explored some of the reasons why individual responsibility for health is deemed to matter morally. We have also touched on some of the possible justifications for treating people differently if they have played a causal role in bringing about an adverse health condition. In the next section, we will examine some of the epidemiologic evidence regarding the social patterning of health behaviors, and canvass some of the criticisms that have been advanced against claims of the type advanced in 4b above.

The Social Patterning of Health Behaviors: Evidence and Ethical Implications

Like health itself, behaviors do not exist in a social vacuum. That is, we have seen how much social structures such as wealth, education, and housing impact health. It would be surprising indeed if behaviors connected to health were somehow immune from the effects of social structures. As noted at the outset of this book, when we think about the social determinants of health, we are taking a birds-eye view to examine health outcomes among and between groups. We are looking at patterns of disease, in part to locate the presence or absence of pronounced differences in

---

[7] Ibid., 115.
[8] Ibid.

important public health indicators. Although these kinds of inequalities are not per se unjust, they are morally significant, and certainly merit investigation and scrutiny, even where we ultimately decide that some such inequalities are fair and just.

There is no reason whatsoever to alter our perspective in examining health behaviors. While such behaviors are significant on the individual level, recall that public health ethics is oriented towards group-level analysis. When we think about health behaviors and their impact on health outcomes, we must similarly look for patterns in those behaviors among and between groups. Thus, epidemiologists and public health scientists tracking and analyzing health behaviors apply similar techniques and approaches to studying behavior as they do to studying disease itself. This makes sense especially where we are attempting to evaluate the connections between behaviors and health outcomes. If such a connection exists, it would make little sense to scrutinize carefully group-level differences in outcomes but only analyze behaviors on the individual level.

Perhaps unsurprisingly, the evidence shows that most risky health behaviors show social gradients that track the social gradient of health. For example, an inverse relationship exists between SEP and adverse health relationships such as "smoking, physical inactivity, less nutritious diets, and excessive alcohol consumption."[9] There is also evidence that unsafe sexual behaviors are linked to SEP and other forms of social disadvantage.[10] One of the obvious questions, of course is why this might be the case. Why is it that more marginalized populations seem at greater risk than more empowered groups of engaging in risky health behaviors?

Jarvis and Wardle engage the question as to cigarette smoking, which represents the single largest cause of preventable death on the planet.[11] They acknowledge, of course, the existence of personal choice regarding smoking, yet argue that too much emphasis on individual agency "fails to address underlying questions of why disadvantaged people are drawn to these behaviours and the nature of the social and individual influences that maintain them."[12] Jarvis and Wardle note a variety of possible explanations for the social gradient in cigarette smoking, including

---

[9] Institute of Medicine (US) Committee on Health and Behavior: Research, Practice, and Policy. Health and Behavior: The Interplay of Biological, Behavioral, and Societal Influences. Washington (DC): National Academies Press (US); 2001. 4, Social Risk Factors. Available from: http://www.ncbi.nlm.nih.gov/books/NBK43750/

[10] Aletha Y. Akers, Melvin R. Muhammad, & Giselle Corbie-Smith, ""When you got nothing to do, you do somebody": A community's perceptions of neighborhood effects on adolescent sexual behaviors," *Social Science & Medicine* 72, no. 1 (2011): 91–99; Pamela J. Bachanas et al., "Predictors of Risky Sexual Behavior in African American Adolescent Girls: Implications for Prevention Interventions," *Journal of Pediatric Psychology* 27, no. 6 (2002): 519–530.

[11] Martin J. Jarvis & Jane Wardle, "Social Patterning of Individual Health Behaviours: The Case of Cigarette Smoking," *in Social Determinants of Health* (2d ed.) (eds. Michael G. Marmot & Richard G. Wilkinson) (2006): 224–237.

[12] Ibid., 225.

- higher rates of smoking initiation;
- stronger perceived rewarding effects;
- higher dependence; and/or
- fewer available coping resources.[13]

Nevertheless, they concede the absence of quality evidence for a causal explanation that integrates these various possibilities. Although there is generally no consensus regarding a primary causal theory that drives the social patterning of behavior, many commentators point out that persistently deleterious living conditions can often create a sense of ennui and hopelessness that seems to increase the likelihood of making risky health decisions. For example, in their study of sexual behaviors in a poor, rural community in northeastern North Carolina, U.S., Akers et al., documented the connections between a lack of neighborhood resources (i.e., adequate recreational options, limited safe environments, etc.) and risky behaviors.[14]

In any event, while much work remains to be done in explaining *why* it is that health behaviors seem to track social gradients (i.e., that less well-off groups tend to engage in riskier health behaviors at greater proportions than more well-off groups), *that* it does track such gradients is generally not disputed. Consider, for example, an ongoing dispute about the relevance of behaviors in producing important population health outcomes connected to the Whitehall Studies, which we discussed in Chap. 3.

The Whitehall investigators were extremely interested in the possible connections between health outcomes and behaviors. In their original results paper, published in 1978, Marmot and his colleagues combined the effects of hypertension, smoking, sedentary lifestyle, and high cholesterol.[15] They found that even the combination of these clinical risk factors did not explain more than a third of the total mortality burden of the social group that experienced the highest such burden (not coincidentally, the group with the lowest social class). Accordingly, Marmot et al. concluded that risky health behaviors were not an especially significant determinant of overall health outcomes.

That such health behaviors exert relatively little effect on outcomes when compared with variables such as class and employment status can be difficult to believe. This is presumably because this evidence undermines the value of the lifestyle model of disease, as well as the focus within dominant traditions of health promotion and education on altering risky health behaviors and increasing salubrious health behaviors. If behaviors are only a minor determinant of population health outcomes, why should we expend significant resources in changing them?

A group of European public health scientists have recently challenged the original investigators' interpretation of the Whitehall data. Using methodological techniques unavailable to the original investigators, Stringhini et al. found that, contrary to Marmot et al's original claims health behaviors accounted for as much as 2/3 of

---

[13] Ibid., 230.

[14] Internal cite.

[15] Internal Cite.

the overall mortality burden of the social group that experienced the highest such burden.[16] In other words, clinical risk factors including risky health behaviors had a substantial role in bringing about the excess mortality experienced by the least well-off group.

Where does this leave us? Obviously, the epidemiologic debate over the true impact of health behaviors cannot be resolved in this chapter or this book. Yet the ethical implications of the social patterning of behavior are crucial, and they connect fundamentally to the social determinants of health.

Ethical Implications and Conclusion

The epidemiologic evidence suggests that health behaviors are in most cases strongly correlated with social and economic conditions. The evidence base typically shows a social gradient of behaviors, with more affluent populations engaging in more salubrious and less adverse health behaviors, and the inverse being true for less well-off populations. But if this is the case, it suggests that there is something about living in adverse conditions that seems to promote unhealthy behaviors. Although the mechanisms for such a connection are not entirely clear, some commentators suggest that a sense of hopelessness is likely one of the factors. If a person, especially a child or youth, imagines their life to lack hope, it is not difficult to imagine why that person would be more likely to engage in riskier health behaviors. Others note that deleterious social conditions may be more likely to diminish a child's self-esteem or retard its development. Lack of self-esteem is strongly linked with unhealthy behaviors, which, again, is a connection that should not be difficult to understand.

Of course, there are always exceptions to the rule. Indeed, in at least one prominent case, the social gradient is reversed in the U.S. The available evidence shows that vaccine-hesitant or vaccine-refusing parents are typically drawn from White, highly educated, and highly affluent communities.[17] Vaccine-refusal—which can almost certainly be deemed a risky health behavior, if not for the individual child, then for the population at large—is correlated with affluence and high SEP, rather than the typical correlation observed between high-risk behaviors and low SEP. But note that even this case is not a counter-example to the idea that health behaviors are socially patterned. Rather, the pattern itself is anomalous—but it nevertheless shows that health behaviors are socially patterned!

For our purposes here, we need not parse out the complex and contested connections between individual responsibility, social contexts, health behaviors, and population health outcomes. It is sufficient to highlight the evidence demonstrating that even where individuals can and do exercise agency in extremely difficult social conditions, those background social conditions exert tremendous effect in shaping

---

[16] Internal Cite.

[17] Jennifer A. Reich, "Neoliberal Mothering & Vaccine Refusal: Imagined Gated Communities & the Privilege of Choice," *Gender & Society* 28, no. 5 (2014): 679–704.

the distribution of health behaviors. Simply put, if a group lives in adverse social conditions, it is more likely that members of that group will engage in riskier health behaviors. This is a group-level phenomenon—it will not be true of every member of the group, and there are significant moral downsides to associating risky health behaviors with individual members of marginalized groups. Specifically, such perceptions, even where rooted in the epidemiologic evidence applying to *groups*, run a significant risk of stigmatizing already marginalized and disadvantaged groups. It should be obvious how intensifying stigma against the least well-off is morally problematic. In addition, perceiving an individual belonging to a marginalized group as fated or destined to engage in high-risk behaviors essentially strips that individual of agency, it disempowers that individual. This is morally problematic because members of disadvantaged groups are already *by virtue of that disadvantage* disempowered. For public health actors to eliminate what little agency such persons may possess seems obviously unethical. It is axiomatic that public health action should empower individuals and communities, not rob them of what agency they enjoy.

These very real risks highlight a conundrum for public health practitioners. On the one hand, the evidence shows that health behaviors are socially patterned and generally track social gradients. On the other, groups are not individuals and some of the most basic ethical obligations of public health practice include the responsibility to empower people and to avoid stigmatizing them. What are public health practitioners to do? The dilemma is real and there is no ethical dictum or principle that can provide a ready answer in any given context. To some extent, this difficulty is simply the challenge posed by evidence-based practice. How does evidence acquired from study of populations apply to the particular person or persons in front of the practitioner? Grappling with the problem is what it means to practice in the health professions. This is not meant as a pleasantry; such a theory of morality derives from one of the great philosophers in Western history: Aristotle himself. He argued that ethics should be thought of in terms of a practice rather than as a set of prior rules and principles, and that the "answers" to difficult moral problems, if they exist at all, are only to be found in the act of trying to resolve them (i.e., by practicing the art of living morally).

Resolving the complex debates and issues that swirl around notions of individual responsibility, health behaviors, and social patterning of such behaviors is beyond the scope of this chapter and of this book. Hopefully, this chapter has mapped some of the conceptual terrain, and identified key points for understanding some of the ethical implications of the debate for public health practice and policy.

Discussion Questions

1. What does it mean to say that health behaviors are socially patterned?
2. How do social and economic conditions influence health behaviors?
3. Does the social patterning of health behaviors matter morally? Why or why not?

4. How much weight should individual responsibility for risky health behaviors have in influencing the scope of obligations we may have to act on the social determinants of health? In other words, given that individuals have some agency for their own health, are we therefore less obligated to intervene on upstream social and economic conditions? Why or why not?
5. Should the facts regarding the social patterning of health behaviors change your practices as a public health professional? Why or why not?

Chapter 5
The Social Determinants of Health & Public Health Practice

After the previous three chapters, it should be difficult to deny both the moral and the public health impact of the social determinants of health. And even while reasonable people of good conscience will disagree on how much of a priority attention to these social determinants are, there is little basis for concluding that such attention is not "a" priority. So, if this is the case, what are the implications for public health practice itself? Of what should it consist? Public health practice is obviously related to public health policy and to public health leadership, but is not identical to either. We might argue that policy related to the social determinants of health is a top priority without necessarily committing any individual public health practitioner to any particular action or intervention. Yet, at the same time, if public health practice does not track priorities in public health policy, such policies are virtually guaranteed to be ineffective.

The evidence of and moral analysis pertaining to the social determinants of health thus presents pressing questions for public health practitioners, both individually and as part of communities of public health practice, whether local, domestic, or global. These include but are not limited to the following:

- Does the evidence regarding the social determinants of health obligate public health practitioners to do anything in particular? Or does it simply recommend certain acts as morally preferable?
- If there are obligations, do they flow to each and every individual practitioner? Or are they social obligations that extend to communities of practitioners as a whole?
- If public health practitioners do have individual moral obligations that flow from the evidence on social determinants of health, where do those obligations rank in relation to other professional obligations?
- If public health practitioners can only implement certain interventions, are those that target social determinants of higher priority than other kinds or categories of interventions?

© The Author(s) 2017
D.S. Goldberg, *Public Health Ethics and the Social Determinants of Health*,
SpringerBriefs in Public Health, https://doi.org/10.1007/978-3-319-51347-8_5

- What is the relationship between public health policy regarding the social determinants of health and public health practice?

This chapter is devoted to exploring these questions and others that relate to the
implications of the evidence regarding social determinants of health for public
health practice.

The Social Determinants of Health and the Proper Scope
of Public Health Practice

Although many guidance documents regarding public health practice and the social
determinants of health seem to take it as a given that we owe collective moral obligations to act on such determinants, it is not clear what the obligations are for individual public health practitioners. After all, given the fallacy of division discussed
in Chap. 2, moving directly from moral analysis of groups to moral analysis of
individuals is unjustified.

Moreover, especially for public health actors working on smaller scales, whether,
for example, as individual practitioners or as part of local health departments, taking
action on the social determinants of health poses very real political risks. Gostin and
Powers note:

> Perhaps the deepest, most persistent critique of public health is that the field has strayed
> beyond its natural boundaries. Instead of focusing solely on narrow interventions for dis-
> crete injuries and diseases, the field has turned its attention to broader health determinants.
> It is when public health strays into the social/political sphere in matters of war, violence,
> poverty, and racism that critics become most upset.[1]

Some commentators argue that while attention to the social determinants of
health is crucial, it does not follow that such attention is wholly or even primarily
the responsibility of public health practitioners. Public health ethicist and lawyer
Mark Rothstein, for example, argues that action on social determinants, ought not
be "annexed" "into the public health domain."[2] Rothstein, a steadfast proponent of
a narrower model of public health practice argues that a broader model that prioritizes action on the social determinants of health will dilute public health to the point
that it becomes "public relations." Broadening public health beyond its focus on
acute public health threats also risks squandering what credibility public health
practice enjoys among policymakers and communities.

Thus, even if the evidence regarding the social determinants of health is taken to
justify some social obligations, it is far from obvious that said obligations are also

---

[1] Lawrence O. Gostin & Madison Powers, "What Does Social Justice Require for the Public's
Health? Public Health Ethics & Policy Imperatives," *Health Affairs* 25, no. 4 (2006): 1053–1060;
quote on p. 1055.

[2] "Rethinking the Meaning of Public Health," *Journal of Law, Medicine & Ethics* 30, no. 2 (2002):
144–149; quote on p. 145.

owed by public health practitioners. Of course, Rothstein's argument is open to the criticism that the narrower model of public health he advocates may have only limited impact on both Relative and Absolute Health:

> [I]f the scope of public health policy under the narrowmodel is not causally related to health, policies consistent with the narrower approach are essentially guaranteed to be ineffective in promoting public health or preventing illness. It is fair to question the utility of public health practices and policies that are expressly intended to avoid addressing or ameliorating the root causes of poor health.[3]

Although this debate rages unresolved, it goes to some of the most fundamental questions in all of public health ethics: what should public health aspire to be? What is most important? What is within the scope of public health? What are important tasks best left elsewhere? These questions underlie what is often referred to as the boundary problem in public health. Namely, once one understands the extent to which social and economic factors affect public health, it seems almost no form of social life is entirely disconnected from health itself. But if this is so, then conceivably virtually any kind of action or intervention could be justified in the name of public health. If virtually everything impacts health, then virtually everything is legitimately within the purview of public health practice. Public health practice itself would know no boundaries.

In turn, a boundless scope for public health raises a number of moral concerns. Because public health practice so frequently involves government authority and state action, a limitless scope of public health imposes few limits on the exercise of government power. The dangers of this are not hypothetical, as public health actors in the modern era have in multiple cases committed troubling acts in the name of public health. For example, in the early twentieth century, many public health leaders and practitioners were active supporters of the U.S. eugenics movement, and played a significant role in the involuntary sterilization of tens of thousands of women in the name of "public health." As historian Alan Kraut as documented at length, immigrants to the U.S. have long been deemed responsible for outbreaks of infectious disease, with stories of mistreatment and abuse at the hands of public health officials being regrettably common.[4]

In short, to claim that serious public health overreach is confined to rare or isolated examples is totally unsupportable based on the ample historical record. It is accordingly no surprise that leading public health law scholars literally define the field as the tension between individual rights and state action in the name of public health.[5] Yet, if the risk of transgressive state action is so real, the boundary problem

---

[3] Daniel S. Goldberg, "In Support of a Broad Model of Public Health: Disparities, Social Epidemiology and Public Health Causation," *Public Health Ethics* 2, no. 1 (2009): 70–83, quote on p. 74.

[4] *Silent Travelers: Germs, Genes, & the Immigrant Menace* (Baltimore: Johns Hopkins University Press, 1994).

[5] Lawrence O. Gostin: Lindsay F. Wiley, Public Health Law Power, Duty, Restrant (Berkeley: University of California Press, 2016).

becomes extremely important for public health ethics: a boundless scope for public health practice is morally concerning, if not downright inadvisable.

All of these concerns leave the public health practitioner concerned with the evidence regarding the social determinants of health in something of a quandary. Given the quality of this evidence base and the virtual consensus among epidemiologists and public health scientists that social and economic conditions are prime determinants of population health, a public health practice that expressly refuses to intervene on such conditions is effectively neutered from the start. The possible impact that public health practice could have on either Absolute or Relative Health (improving overall population health or compressing health inequalities) would be substantially limited by an *ex ante* decision to implement public health interventions that are targeted at downstream variables rather than at root social determinants of health. From a moral standpoint, this seems at the very least to be suboptimal and might well be unacceptable.

On the other hand, public health practice that is wholly targeted at changing upstream social and economic conditions may be less likely to be effective at all, may squander what political and public legitimacy public health enjoys, and also runs a substantial risk of facilitating government overreach in the name of public health. How is the public health practitioner to resolve this dilemma?

There is likely no neat and easy "solution" to this problem, as it goes to deep and fundamental questions about what the community of public health practitioners aim to "do and be." What is at stake is no less than the collective identity of public health practitioners, and there is no formula or algorithm that will dictate a plain answer to these questions. Nevertheless, it does not follow that guidance of any kind is impossible and that no plausible paths exist. Indeed, regardless of the difficult quandary of action posed by the evidence regarding the social determinants of health, there is a near consensus among many public health stakeholders and leaders that total inaction on the social determinants of health is simply not an option. Consequently, there are no shortage of programs and efforts driven by public health practitioners intended to ameliorate adverse social and economic conditions. The following section examines some of them.

Examples of Programs that Integrate Action on the Social Determinants of Health into Public Health Practice

Programs that aim to act on the social determinants of health often have to adjust the level of intervention. One way of explaining this, and the shift in insight that is required, is by discussing one of key concepts in all of public health: prevention.

There is no doubt that prevention is fundamental to public health. In fact, most textbooks and public health leaders situate a focus on prevention as one of the defining features of the field itself, and one that distinguishes it from other health professions. (While of course most if not all fields in the health professions properly maintain some interest in prevention, it is absolutely central to public health and defines the field in a way that differs from its role in medicine, for example). There are at least three principal levels of prevention: primary, secondary, and tertiary. Primary prevention emphasizes risk reduction as a way of barring the onset of disease itself. An example here would be an intervention designed to encourage exercise so as to reduce the risk of developing type II diabetes or coronary artery disease. Secondary prevention "attempts to identify a disease at its earliest stage so that prompt and appropriate management can be initiated."[6] Screening programs are classic examples of secondary prevention. Tertiary prevention "focuses on reducing or minimizing the consequences of a disease once it has developed."[7] Most forms of medical care, as CDC notes, are actually a form of tertiary prevention.

Note that the levels of intervention that correspond to these different forms of prevention are all located downstream the causal pathway. Tertiary prevention occurs after disease processes are already impacting the health of the patient. Secondary prevention occurs either right before or right after the onset of disease. Even efforts at primary prevention, which emphasizes preventing the onset of disease, has traditionally focused proximal to the onset of disease (by identifying at-risk groups who may already be at substantially increased risk of developing a particular disease in a short-to-intermediate time horizon).

Some public health leaders have called for a fourth form of prevention, one that moves the level of intervention higher up the causal chain: primordial prevention. Epidemiologist John Last notes that

> [p]rimordial prevention…aspires to establish and maintain conditions to minimize hazards to health…it consists of actions and measures that inhibit the emergence and establishment of environmental, economic, social and behavioral conditions, cultural patterns of living known to increase the risk of disease.[8]

Primordial prevention efforts are targeted high up the causal pathway of disease, reaching up beyond lifestyle and behavioral modifications to upstream determinants of health. David Kindig argues that

> the more classic definition of prevention is too rooted in the lifestyle modification efforts of the past 40 years so that equal attention is not given to the upstream social determinants—or that it leads to taking the comfortable position that since improving income and education is so difficult we leave it to others (such as letting the Treasury and Federal Reserve worry about unemployment for us).[9]

Kindig, then, expressly rejects the narrow model of public health that argues for leaving collective action on social and economic conditions to entities and branches of government beyond traditional public health agencies and actors. But what specifically do interventions focused upstream look like?

---

[6] CDC, http://www.cdc.gov/arthritis/temp/pilots-201208/pilot1/online/arthritis-challenge/03-Prevention/concept.htm

[7] Ibid.

[8] David Kindig, "Have You Heard of 'Primordial Prevention?'" available at http://www.improving-populationhealth.org/blog/2011/05/primordial_prevention.html

[9] Ibid.

One example comes from Alameda County, California, where the County Public Health Department initiated a "Place Matters" initiative that reflects a focus on the ways in which multiple social determinants converge in the concept of "place" to shape health outcomes (SEP, housing, racial segregation, environmental hazards, exposures to violence and/or safe recreational areas, etc. are all part and parcel of "place").[10] The Health Department hired a full time "Place Matters" coordinator, charged the planning team with a needs assessment focusing on "historical and current policies and practices at the root of inequitable community conditions."[11] The team emphasized six key social determinants (criminal justice, economics, education, housing, land use, and transportation).

Following their needs assessment, the planning team engaged the local community and collaborated to create a local policy agenda identifying priorities for action on the social determinants that the community identified as most critical. Specific engagements included successful efforts

- preventing a local utility from shutting off water to local tenants during the national recession in 2009;
- altering requirements for rental inspections to enhance code enforcement that regulates the conditions under which low-income and marginalized communities disproportionately reside;
- creating a module within truancy hearings that address and remedy school absenteeism that is caused by chronic illness such as asthma.[12]

All of these programs are examples of public health interventions implemented at the local level, via community engagement, that target upstream social determinants of health.

In another case, a group from the Bloomberg School of Public Health at the Johns Hopkins University focused on the city of Baltimore's efforts to rewrite the zoning code.[13] Zoning can impact population health in a number of ways, by shaping patterns of racial segregation, environmental hazards, transportation, and recreation (among others). The investigators conducted in-depth interviews with key stakeholders and conducted a health impact assessment ("HIA") with particular attention to the ways in which zoning shaped major social determinants of health. Assessment of the impact of the HIA was ongoing, but the project itself was unmistakably geared towards informing key actors with the knowledge needed to intervene on the upstream social determinants most relevant to the local community's concerns.

---

[10] Katherin Schaff et al., "Addressing the Social Determinants of Health through the Alameda County, California, Place Matters Policy Initiative," *Public Health Reports* 128 (Supp. 3) (2013): 48–53.

[11] Ibid., 128.

[12] Ibid.

[13] Rachel L. Johnson Thornton et al., "Achieving a Healthy Zoning Policy in Baltimore: Results of a Health Impact Assessment of the TransForm Baltimore Zoning Code Rewrite," *Public Health Reports* 128 (Supp 3) (2013): 87–103.

Although there are, fortunately, no shortage of specific examples of projects such as those described above, it is also worth noting programmatic approaches to integration of the social determinants of health into public health practice. One of these is the approach known as "Health in All Policies," or "HiAP." The point of departure for HiAP is a concept long trumpeted by stakeholders committed to action on the social determinants of health: social policy is health policy. That is, the evidence is clear that policies and practices within a variety of sectors (e.g., housing, labor, education, etc.) exert a profound impact on health across the lifespan. Thus, limiting evaluation of the health impact of policies to those policies arising expressly from within the health sector is ill-advised if the goal is to maximize impact on overall population health (Absolute) and on the compression of health inequalities (Relative).

HiAP is geared towards this problem, and focuses on leveraging health expertise to sectors well outside the traditional channels of public health and health care policy. Thus, policymakers working on particular housing laws or tax laws or education laws will be educated on the likely downstream health impact of the proposed laws and policies. The ultimate goal of HiAP is to integrate attention to the social determinants of health across the entire policy domain, at local, state, national, and international levels. Public health scientists and stakeholders are currently working on models to test the impact of HiAP approaches itself.[14]

Another programmatic response to the evidence regarding the social determinants of health are medical-legal partnerships ("MLPs"). MLPs represent an effort to integrate medical and legal services to better serve the social needs of patients. At its core, MLPs are unquestionably driven by a focus on the social determinants of health. The justification for MLPs is precisely that a huge amount of the health problems with which people present in clinical settings are driven by social and economic conditions. In some cases, trained attorneys may be able to assist patients in finding the resources needed to ameliorate adverse social and economic conditions in ways that health care providers cannot. Thus, for example, a family living in substandard housing may expose young children with asthma to damp dwellings with high allergen exposure, including mold and other potentially severe risk factors. An attorney working within a defined MLP may be able to assist the family in initiating legal action that would force the landlord to remedy the dwelling as mandated under applicable laws and ordinances. These legal efforts can have a substantial impact on the overall health of patient, family, or even potentially a community. Moreover, because social disadvantage tends to cluster, MLPs can potentially exert a larger impact on poorer communities than on affluent ones. In such cases, MLPs might help *compress* the health inequalities between rich and poor, thereby satisfying the Relative Health prong.

However, although there are case studies and observational reports indicating some health impact for MLPs, the studies thus far lack the controls and rigor needed

---

[14] Fran Baum et al., "Evaluation of Health in All Policies: Concept, Theory and Application," *Health Promotion International* 29 (Supp 1) (2014): i130–i142; Adrian E. Bauman, Lesley King, Don Nutbeam, "Rethinking the Evaluation & Measurement of Health in All Policies," *Health Promotion International* 29 (Supp 1) (2014): i143–i151.

to fully evaluate the ultimate health impact of MLPs.[15] Akin to HiAP, further data is needed to corroborate the promising preliminary signs regarding the health impact of MLPs. Nevertheless, both of these approaches represent programmatic attempts to act on the social determinants of health via policy and practice.

Conclusion

One of the criteria by which many judge ethical frameworks is the extent to which they are able to guide action. To be sure, human action is messy and complicated, and ethical frameworks are not formulas or algorithms that spit out the "right" answers to difficult moral problems. Nevertheless, if as a society we owe any moral obligations to act on the social determinants of health, it is plausible to assert that such obligations should in some way be translated into public health practice.

However, as discussed above, the existence of some collective obligation to act on the social determinants of health does not obviously imply any specific obligations for individuals within a society. Nor does it automatically follow that any such action is the primary responsibility of public health actors. In spite of these complexities, most public health leaders and guidance documents expressly and willingly affirm a responsibility to act on the social determinants of health. As a result, public health practitioners will likely encounter programs, policies, and procedures related to such action, and it is important to think carefully about the rationale for such acts, how those rationales map onto issues of priority-setting in public health practice, and the extent to which ensuing interventions are connected to the epidemiologic evidence regarding social determinants of health.

Discussion Questions

1. Do you think individual public health practitioners have a moral obligation to act on the social determinants of health? Why or why not?
2. How much, if any, of its limited resources should a local health department allocate towards action on the social determinants of health? Justify your allocation decision.
3. Provide two examples of ways in which individual or community-based public health practitioners can act on social determinants of health.
4. How likely is it that an HiAP approach can effectively change upstream social and economic conditions? Explain your answer.

---

[15] Tishra Beeson, Brittany Dawn McAllister, & Marsha Regenstein, *Making the Case for Medical-Legal Partnerships: A Review of the Evidence* (National Center for Medical Legal Partnerships: February 2013), http://medical-legalpartnership.org/wp-content/uploads/2014/03/Medical-Legal-Partnership-Literature-Review-February-2013.pdf

Chapter 6
Conclusion

The overall aim of this brief book has been to explore some of the ethical implications of the evidence regarding the social determinants of health for public health practice and policy. A central question for all of public health ethics asks 'what are the paramount goals for public health itself?' These questions frame any moral analysis of the role that evidence and action on the social determinants of health ought to play. Tracking a number of public health scientists and ethicists, this book has adopted two objectives as normatively paramount for public health: improvements in overall population health (Absolute Health) and compression in health inequalities (Relative Health). Either, absent the other, is ethically suboptimal. Exclusive focus on improvements in overall population health makes very likely an expansion in population health inequalities that often tends to track social fault lines such as class, gender, and race. In such cases, larger gains in health status among the affluent are met by much smaller increases in health among the most deprived. Although not all health inequalities or increases in health inequalities are unjust, it is plausible to suggest that many of them are inasmuch as they are rooted in background social conditions and structures that at least intuitively strike many as unfair. (Obviously, these intuitions are no substitute for careful and thorough analysis, but for reasons discussed in Chap. 4, such intuitions are nevertheless legitimate clues as to the presence of injustice in public health policy and practice).

On the other hand, we could compress health inequalities simply by reducing the health status of the most well-off. This is known as leveling down, and while it might well be legitimate to redistribute income or wealth of the most affluent, few would agree to directly reducing the health of the most well-off as a means of narrowing the health gap between the affluent and the deprived. This argument suggests that while improvements in Relative Health via compression of health inequalities is an important goal for public health, untethered to any effort to improve population health for all, it is morally incomplete or even inappropriate.

If Absolute Health and Relative Health are properly deemed two primary goals of public health, the significance of action on the social determinants of health

© The Author(s) 2017   51
D.S. Goldberg, *Public Health Ethics and the Social Determinants of Health*,
SpringerBriefs in Public Health, https://doi.org/10.1007/978-3-319-51347-8_6

becomes clear after an examination of its evidence base. That is, the evidence is overwhelming that the prime determinants of health and its distribution in human populations are the social and economic conditions in which people work and live. It is difficult to imagine what public health actors aim to do if improving Absolute and Relative Health are not principal among their objectives in both practice and policy. Support for this flows from various frameworks of justice that ground public health action. The theory explored in most detail herein is Powers and Faden's health sufficiency model of social justice. It integrates attention to both Absolute and Relative Health, and takes the problem of systematic disadvantage as core to assessing the moral obligations that ought to guide public health practice and policy.

Despite the considerable guidance that rigorous frameworks of justice offer, this brief book has endeavored to complicate the ethical analysis. It is not obvious exactly what moral obligations flow from a generic commitment to act on the social determinants of health, nor is it plain who owes them (Groups? Individuals? Which ones of each?). Moreover, while the evidence on social determinants of health shows that a social gradient in health outcomes is tracked closely by a social gradient in health behaviors, deciding the extent to which individual responsibility for health mitigates any collective responsibilities to act on social determinants is also difficult. That is, although there is active debate on the subject, there is nevertheless substantial evidence that health behaviors mediate health outcomes. So, inequalities in health outcomes are correlated with inequalities in healthy/risky health behaviors. In other words, health behaviors are socially patterned.

However, the ethical implications of this fact remain far from clear. Even individuals living in adverse social and economic conditions likely retain some measure of agency, some capacity to engage in healthier or riskier health behaviors. Even if not all of our behavior is entirely within our control, some of it remains so, even where social and economic conditions work to undermine individual agency. How does this core of individual agency for health affect any collective moral obligations we have to act on the social determinants of health? This is a complex question beyond the scope of this brief book, but it is one that is impossible to even begin to answer without a sufficient understanding of the social determinants of health and the social patterning of health behaviors.

Finally, it is important to think carefully about the implications of the evidence on the social determinants of health for public health practice. Even this inquiry is more complex than it might initially seem. Some public health ethicists argue that even if we owe some kind of social obligation to act on the social determinants of health, it is far from clear that it is public health practitioners who must satisfy that obligation. In fact, it might be the case that public health actors should not be the group charged with ameliorating adverse social and economic conditions, as such action is likely to be politically and socially controversial, may squander what credibility public health as an enterprise enjoys, and carries with it substantial risk of government overreach onto individual liberties. There is a substantial history of state transgression on individual rights in the name of public health, which makes this last concern worth taking seriously.

On the other hand, a claim that public health actors have no moral obligation to act on the primary determinants of population health outcomes seems to effectively neuter public health action. What good is public health practice if by definition it is unlikely to have much impact on important population health outcomes? There are very real questions that exist regarding the scope of the obligations to act on the social determinants of health that public health actors may owe. Nevertheless, given mandates of justice, it is implausible to argue that public health actors owe no duties of any kind to intervene on the social determinants of health.

Index

© The Author(s) 2017
D.S. Goldberg, *Public Health Ethics and the Social Determinants of Health*,
SpringerBriefs in Public Health, https://doi.org/10.1007/978-3-319-51347-8

Printed in the United States
By Bookmasters